Sustainable Energy Transitions

Dustin Mulvaney

Sustainable Energy Transitions

Socio-Ecological Dimensions of Decarbonization

palgrave
macmillan

Dustin Mulvaney
San José State University
San José, CA, USA

ISBN 978-3-030-48911-3 ISBN 978-3-030-48912-0 (eBook)
https://doi.org/10.1007/978-3-030-48912-0

Cover Design Specs: Tom Howey

This Palgrave Macmillan imprint is published by the registered company Springer Nature Switzerland AG
The registered company address is: Gewerbestrasse 11, 6330 Cham, Switzerland

Acknowledgments

This book is dedicated to all my teachers, and our orange cat César (2002–2020). There are a bunch of people to thank for helping me be a better teacher. I took courses from some wonderful instructors, in high school, college, and graduate school. In college, Professors Angelo J. Perna, R.P. Thompkins, and Trevor Tyson were major influences. Some of my teaching mentors include Professors Melanie Dupuis, David Goodman, Margaret FitzSimmons, Eric Katz, Ronnie Lipschutz, Roberto Sanchez-Rodriguez, Donna Haraway, Nancy Jackson, Ali Shakouri, and Ben Crow, all of whom I was a teaching assistant for, or who I took classes from. I learned and continue to learn a tremendous amount from my fellow graduate students in Environmental Studies at University of California, Santa Cruz (UCSC). Thanks to Daniel Press, who hired me to develop and teach my first energy course at UCSC in 2008. I had an amazing teaching-focused interdisciplinary postdoctoral experience in Electrical Engineering and at College 8 (the Rachel Carson College) at UCSC where I really began to refine pedagogy. I owe a debt of gratitude to the Switzer Foundation and the Silicon Valley Toxics Coalition, two organizations that helped me commit time to learn about this industry. Many thanks also to Professor Alastair Iles at the University of California, Berkeley, who became a critical mentor to me very early as I developed research questions around energy transition. This textbook is intended for my sustainable energy strategies course at San José State University (SJSU), where there is a deep commitment to sustainable energy in the Environmental Studies Department, Frank Schiavo, dating back to the early 1970s. The enthusiasm of the students at SJSU is a major motivation for this work. My colleagues in Environmental Studies also have been very supportive—Professors Lynne Trulio, Rachel O'Malley, Gary Klee, Costanza Rampini, Will Russell, Bruce Olzewski, Katherine Cushing, Beniot Delaveau, Terry Trumbell, and Pat Ferraro.

There is also a vibrant community of people studying energy transitions that deserves acknowledgment and thanks. There is a great teaching resource I have used for quite a few years hosted by Chris Nelder and produced by Justin Ritchie; it's a podcast called *The Energy Transition Show*. It is a wonderful resource for students, and I recommend trying to integrate it into lesson plans around this book. Another similar set of resources were developed over the years by Stephen Lacey, starting way back with *Inside Energy News* and now *The Energy Gang* and *The Interchange* shows, which I also learned much from. Finally, our professional associations and graduate school colleagues, and Twitter friends, are so important for shaping the way we think so I want to thank a lot of friends from over the years—Jill Harrison, Max Boykoff, Mark Buckley, Jill Harrison, Roopali Phadke, Matt Huber, Jenn Bernstein, Damian White, Timmons Roberts, Costa Samaras, Hanjiro Ambrose, Jason Douglas, Paul Robbins, and Morgan Bazilian. Great work all around people. We can make sustainable energy happen.

About This Book

Systems that produce, deliver, and consume energy all around us are undergoing a transition. This is a textbook that I hope reaches people interested in learning about the socio-ecological dimensions of energy system transitions from multiple disciplinary perspectives, including ideas and concepts from engineering, economics, and life-cycle assessment to sociology, political science, anthropology, policy studies, the humanities, arts, and some interdisciplinary thinkers that defy categories. One prominent voice in current debates about energy transitions are argue to act on decarbonizing energy systems to mitigate climate impacts from carbon pollution from energy supplies. But other socio-ecological systems will be transformed and may benefit from shifts in energy use and production patterns. In 2020, 80% of global energy is still supplied from fossil fuels. Many places have taken great strides toward decarbonizing some aspects of life in 2020, but there are many miles to go to make a sustainable future.

The adjective "socio-ecological" refers to the set of human and nonhuman systems interweaving the biophysical world and its ecologies with the metabolism of human civilization. Socio-ecological systems tied to our energy use are complex and often across great geographical distances, so the book aims to draw case studies from around the world to bring into perspective the various ways that human ingenuity is working to provide renewable and clean energy, and tackling its side effects.

The multiple disciplines presented in this textbook aim to build bridges across the social and natural sciences and humanities to introduce readers to the development of energy and efforts and prospects of an energy transition. I designed this book with students with a sustainable energy focus in mind, borrowing from our concentrations in energy in Environmental Studies at SJSU. I have integrated case studies, figures and tables, exercise problem sets, pictures and diagrams of different energy systems, and links to further resources for further exploration of energy questions. In my own courses, I aim to use this book as the bedrock for a project-oriented course, with lots of class discussions and group projects. But it can be read and taught page by page at whatever pace. The book is also written in a way that can be structured around slides and conventional lectures. The lecture slides that accompany these chapters are available on the web.

Contents

About the Author

Dustin Mulvaney

is a Professor in the Environmental Studies Department at San José State University (SJSU). He received his doctoral degree from the Environmental Studies Department at the University of California, Santa Cruz, and obtained a Master of Science in Environmental Policy Studies and Bachelor of Science degree in Chemical Engineering from the New Jersey Institute of Technology. He started at SJSU after an appointment at the University of California, Berkeley, where he was a National Science Foundation Science and Technology Studies Postdoctoral Scholar in the Department of Environmental Science, Policy, and Management. His contributions to research include work on environmental justice, solar energy commodity chains, and public lands and energy development in the American West. His book *Solar Power: Innovation, Sustainability, Environmental Justice*, published by the University California Press in April 2019, reveals extensive research on photovoltaic sustainability, chemical stewardship, and recycling management, including participation in the development of the e-Stewards recycling standard for photovoltaics with the Basel Action Network and as a member of the Joint Committee for a sustainability leadership standard for photovoltaics at the National Standards Foundation International. He previously worked for a bioremediation start-up as a project engineer, designing cleanup strategies for soil and water contamination. Prior to that, he was a chemical process engineer for a Fortune 500 chemical manufacturer.

Abbreviations

AC	Alternating current
ACEEE	American Council for an Energy-Efficient Economy
AI	Artificial intelligence
AV	Autonomous vehicles
BAU	Business-as-usual
BEV	Battery electric vehicle
BWR	Boiling water reactors
CAFE	Corporate average fuel economy
CARB	California Air Resource Board
CCS	Carbon capture and sequestration
CDC	U.S. Centers for Disease Control
CdTe	Cadmium Telluride
CEC	California Energy Commission
CIGS	Copper indium gallium diselenide
CITES	Convention on International Trade in Endangered Species of Wild Fauna and Flora
CPP	Critical peak pricing
CPUC	California Public Utilities Commission
CSR	Corporate social responsibility
DC	Direct current
DG	Distributed generation
DLC	Direct load control
DOE	Department of Energy
DR	Demand response
EE	Energy efficiency
EGS	Enhanced geothermal systems
EIA	U.S. Energy Information Administration
EJ	Environmental Justice
EPAct	U.S. Energy Policy Act
EPR	Extended producer responsibility
EROI	Energy Return on Investment
EV	Electric vehicle
FAME	Fatty Acid Methyl Esters
FAO	United Nations Food and Agriculture Organization
FERC	U.S. Federal Energy Regulatory Commission
FIT	Feed-in-tariff
GHG	Greenhouse gases
GRI	Global Reporting Initiative
HFC	Hydrofluorocarbon
HVAC	Heating, ventilation, air-conditioning
ICE	Internal combustion engine
IEA	International Energy Agency
IEEE	Institute of Electrical and Electronics Engineers
INDC	Intended National Determined Contributions
IoT	Internet of things
ISO	Independent system operator
LCA	Life-cycle assessment
LCFS	Low-carbon fuel standard
Li-ion	Lithium ion

LNG	Liquefied natural gas
LWR	Light water reactors
Mox	Mixed oxide fuel
MSW	Municipal solid waste
NHTSA	U.S. National Highway and Transportation Safety Administration
NREL	National Renewable Energy Lab
OPEC	Organization of Petroleum Exporting Countries
OTEC	Ocean thermal energy conversion
PURPA	U.S. Public Utilities Regulation Policy Act
PV	Photovoltaics
RCP	Representative Concentration Pathways
REDD	Reduced Emissions from Deforestation and Degradation
RoHS	Restriction on Hazardous Substances
RTP	Real-time prices
SDG	Sustainable Development Goal
STS	Science and technology studies
TOU	Time of use
TSS	Total suspended solids
U.S. EPA	U.S. Environmental Protection Agency
UIC	Underground injection control
UNCED	United Nations Conference on Environment and Development
UNFCCC	United Nations Framework Convention on Climate Change
V2G	Vehicle-to-grid
VOC	Volatile organic compound
WEC	Wave energy converter
WTW	Well-to-wheels
WWS	Wind, water, sunlight strategy
WWTF	Wastewater treatment facility

List of Figures

Energy Transitions

Contents

D. Mulvaney, *Sustainable Energy Transitions*,
https://doi.org/10.1007/978-3-030-48912-0_1

Learning Goals

By the end of this chapter, readers will have:

- a broad understanding of historical patterns of energy transitions;
- an introduction to theoretical frameworks and debates about energy transitions;
- an appreciation for how histories of energy transition manifest today, and
- a roadmap for the rest of the book.

Overview

The way energy is extracted, transported, used, and disposed of is changing right before our eyes, again. Many say we are witnessing a major energy transition to renewable energy, or maybe several energy transitions at the same time. It is not the first time. Energy transitions recur throughout human civilization. For much of human history, energy access, resource development, and the accompanying technological changes improved quality of life, albeit some more than others, for most of the human population. Putting biomass energy to use led to sedentary human communities, agriculture, metallurgy, and so on; it powered most of human history. By the mid-nineteenth century, fossil fuels provided many of the modern conveniences of life—food, electricity, transportation, cities, and industries.

Widespread social and environmental damages and concerns about climate change raise questions about the wisdom of using polluting energy sources for light, heat, and motion. How could we hasten an energy transition toward providing access to those in energy poverty and decarbonization conditions—making low-carbon energy systems with the fewest environmental impacts? How can we do this in a way that reflects concerns for labor, waste, air pollution, soils, fish, wildlife, and water use? This chapter introduces readers to the key concepts, unanswered questions, and critical debates about energy transitions.

Definition

Energy transitions are socio-technical processes that reshape the nature or patterns of use of energy resources and/or technologies.

1.1 Energy Transitions and the Anthropocene

The environmental crises and global changes that characterize the Anthropocene—climate change, biodiversity loss, air pollution, water use and pollution, deforestation, nitrification, pollution and habitat loss in our oceans, and soil quality degradation, to name a few—have spurred interest in the socio-ecological

1

Energy Use and the Coronavirus Pandemic

The novel coronavirus pandemic revealed many immediate changes related to global public health measures such as shelter-in-place and social distancing orders that severely curtailed normal economic activity. Satellite data presented by the National Aeronautics and Space Agency (NASA) and the European Space Agency on air pollution revealed steep declines from reduced transportation and smog pollution in the short term. What happened next? The moment provides an opportunity to think about the interconnectedness of energy systems on several time scales. Falling gasoline consumption has impacted communities that produce ethanol and petroleum. Supply chains for electric vehicles, photovoltaics (PVs), and batteries all have faced disruptions. Then there are questions about the response to repair the economy. Will there be investments in new housing, clean energy infrastructure, and green jobs? The coronavirus-caused COVID-19 pandemic could have lasting implications for energy transitions.

connections that link human activities to natural resource use and environmental change. The question so many are grappling with is how to shift economies and societies toward more sustainable use of natural resources and land, while minimizing waste and pollution. Questions about air and water pollution have long been linked to energy systems, as have questions about hastening the evolution of energy systems and access to them more generally. Increased attention to questions of energy and climate have taken on great urgency in recent years, owing to concerns about exceeding climate budgets and talk of runway warming scenarios for our planet—a hothouse Earth (Steffen et al. 2018). Research on the social lives that experience the extraction, transportation, and generation of energy generally find improvements in quality of life for the many benefits energy brings us (Smil 2010). Energy infrastructure and extraction can also be experienced as environmental degradation, worker exploitation, and other environmental inequalities, links that are sometimes masked because of the multiple nodes connecting consumers to natural resources. Today, human influences on the Earth's ecosystem and processes are more apparent. Narratives about the undisputed benefits of mass consumption have shifted from an unquestioned modernist ideal to a major driver of global environmental change (Adger et al. 2001).

Definition

A socio-ecological system describes human and Earth-system interactions as dynamic, interconnected, and co-produced by nature and society.

How do social processes cause environmental change? Can green production be achieved by technology to make stuff with fewer impacts? Or is it about how we act? Should we consume less of what we do today? These are key questions facing civilization as it thinks about sustainability—the idea that human impacts on ecosystems can be minimized and natural resources be made available for future generations. Dematerialization and sustainable development are two competing frames to displace consumption as the modernist ideal of consumption, meaning "what it takes to live and thrive" (Yates 2012). Early reports focused on environmental resources include *The Limits to Growth* (Meadows et al. 1972) and *Our Common Future* (Brundtland 1987). These reports legitimated sustainable development as a global concern and set out frameworks to advance an international dialogue among nations on environmental responsibility, poverty alleviation, and planetary stewardship. International frameworks to manage, or at the very least evaluate, the multiple social and environmental challenges emerging from this early work included the Convention on International Trade in Endangered Species of Wild Fauna and Flora (CITES), Intergovernmental Panel on Climate Change (IPCC), the Montreal Protocol to protect the ozone layer, *Agenda 21*, and the Sustainable Development Goals (SDGs).

Thinking about Coronavirus

2020 saw the start of the novel coronavirus pandemic, causing severe disruption to energy systems. Reflect on the changes that you noticed in your everyday routines during the global public health measures. What images related to energy do you remember from television, news, or social media? How did the public health measures affect your own patterns of energy use? What were the short-term implications for energy industries? How might it shape our relationship to energy in the long term?

There are a number of ways that the concept of energy transitions is used. In the 1970s, the phrase was mainly used to describe the challenge of hastening energy access to the parts of the world in energy poverty, moving households and communities up the energy ladder to higher quality energy that would improve the relative well-being of those populations (Leach 1992). Around this time it was also used to describe the growing demand for coal resources in the rapidly growing cities of the southwestern United States. Today, the phrase "energy transition" is used to describe the move toward a low-carbon economy (Geels et al. 2017). Acknowledging the need for a *just transition* refers to how to design and the process of decision-making that results in an energy system that benefits all, while remaining within environmental, natural resource, or economic constraints. Social scientists and technological futurists explain why social transitions happen or do not take hold with nearly one hundred theories on social change according to energy and technology scholars Ben Sovacool (2013) and David Hess (Ryghaug et al. 2018; Hess 2012).

Characteristics of the Anthropocene

Humans are a dominant global force on planet Earth. The extent to which the period deserves a name is a matter of debate among geologists and anthropologists. One somewhat academic question about the Anthropocene is where to place the so-called golden spike—to mark the transition to this human-dominated geologic time from the Holocene, the period that defines the past 12,000 years of geology, to one dominated by humans. Some golden spike contenders include the first major imprints on the atmosphere from fossil fuels—emissions of methane and carbon dioxide—which have clear anthropogenically derived signals in isotopes in the balance of carbon 12 to carbon 14. The date that marks the start of atmospheric nuclear testing—the plutonium 239 line layered down in the early 1950s—is currently another leading candidate.

Other "-cenes" have emerged as well, such as the "plasto-cene" (Langlois 2018), "petro-cene," "capital-o-cene" (Moore 2015), "chthulucene" (Haraway 2015), "pyrocene" (Pyne 2015), "Anthropo-obScene" (Swyngedouw & Ernstson 2018), and the "its-too-late-o-cene" (White 2019). Perhaps we will see the time of the 2019–2021 pandemic as the viro-cene. A few other contenders for the marker of human influence include large-scale deforestation, major anthropogenic shifts in patterns of soil erosion, extensive wetland reclamation for global rice production, and crop domestication—all of these activities left permanent marks on the history of Earth, clearly caused by humans. Incidentally, many of these transitions began long before the fossil fuel period. Some scientists argue that rather than try to establish a benchmark line to define the start of the Anthropocene, it would be better to have an informal use depending on the context (Ruddiman 2018).

Energy transitions can be represented by significant changes or divergences from the "business-as-usual" (BAU) trends for energy use. Energy historian and interdisciplinary researcher Vaclav Smil calls energy transitions fundamental processes in the evolution of societies, and underscores their glacial pace, pointing to key transformations in prime movers like the steam turbines. There are many different projections for energy use by 2050; the International Energy Agency (IEA) (2010) suggests power demand will increase from 18 trillion watts in 2020 to somewhere between 25 and 30 trillion watts by 2050. Yet, things are changing. Starting in 2015, the world started to build more renewable energy than energy infrastructure to burn fossil energy (Platts 2017).

Anticipating what our energy future looks like is the task of energy modelers and analysts, but also designers, social scientists, ethicists, artists, and a whole number of people whose ideas, creativity, and expertise could significantly influence it substance. While colloquially the terms forecast and projection are used interchangeably, there is a critical distinction between the two. Projections are trends taken into the future, based on existing trends or some BAU scenario. Forecasts are made by taking these projections and modifying them with assumptions about the future that might change rates of change or adoption of a technology. These might include assumptions about driving forces—new market developments, regulations, innovations, demands for new stuff, and/or behavioral changes—that are pushing one sector of technologies, like renewables, over others that will shift energy use in the future.

To begin to understand our energy systems is helpful to distinguish between primary energy resources and final fuel products, energy carriers, and energy end uses.

Definition

Primary energy sources are the natural resources taken from the Earth: coal, "wet" natural gas (wet because it contains water, methane, ethane, and other gases), petroleum, solar and wind power, uranium, and other direct sources of energy harvested.

Primary energy sources are converted, transported, and made into final fuel products and energy carriers, which provide energy services like light, heat, and motion. Final fuel products include gasoline, "dry" natural gas (dry because it mostly contains methane), wood for a stove or campfire, hydrogen, and electricity. Energy carriers include electricity, hydrogen, and steam.

Definition

Final fuel products and energy carriers are the energy resources that directly provide energy services.

The unsustainable use of natural resources is the critical consideration in the project of energy futures. Dialogues on how to these never resolved the inherent contradictions in the concept of sustainable development, such as whether degrowth was needed to fully satisfy sustainability goals, or to develop, sustainably. *Agenda 21*—a United Nations program—intended to help developing countries tackle poverty and environmental issues, opened up connections between concepts in sustainability and to the managerial discourse and business of corporate social responsibility (CSR) (Sadler & Lloyd 2009). CSR is an approach to sustainability that emphasizes the triple bottom line, with a theory of social change that sees the goal as encouraging private sector responses to market-based approaches (Milne & Gray 2012). Voluntary standards, industry benchmarks, and environmental and social disclosures favored by CSR make some of these spaces governable by getting companies to respond to concerns from shareholders or about reputational risk. Resolving questions of what is sustainable development, and how to foster it in energy transitions, will continue to be a critical conversation.

Amory Lovins (1976), E.F. Schumacher (1973), and Vaclav Smil (2004) are just a few of the early writers on energy transition, all providing big-picture views over the long arc of energy in human history. But as we will see, the topic of energy systems and change is much broader. Through the work of Elizabeth Shove, we learn about the social construction of comfort and how influences on energy use are shaped by expectations of comfort. Energy transitions are also represented by social movements that are motivated more by climate change, innovation policy, racial justice, and the creation of green jobs. Wind, water, and sunlight (WWS) strategies are blueprints to replace energy systems with one run entirely on electrification and renewables (Jacobson & Delucchi 2011). Another team developed a so-called solar grand plan, a plan to harvest solar energy from the US southwest in a major electricity energy transition (Zweibel et al. 2008). Some entrepreneurs proposed an ill-fated plan called DESERTEC to power Europe with solar power from North Africa. Many critics noted with skepticism the irony of exporting electricity away from places that need it most.

> Geographer Martin Pasqualletti describes energy problems as social problems with a technical component, not the other way around like it tends to be treated.

Geographer Scott Jiusto (2009) notes that the "paradox of conventional energy systems is that they are, at present, essential to economic productivity and social well-being and yet enormously destructive, crisis-prone and unsustainable" (p. 547). Energy transitions can be conceived of at several scales and across several different domains of energy services. Shifting the current global energy system from fossil fuels to one powered on biofuels and electricity from intermittent renewables could be "necessarily a prolonged, multi-decadal process" (Smil 2010). Specific energy resources, or devices that use energy, turnover relatively quickly so long as they are available globally.

Today, energy supplies available are diverse as ever, and, at the same time, changes to the larger energy system may move at a glacial pace. If energy transitions are really social changes, there are numerous social theories that offer insights. It may be appropriate to consider social movement theory, institutional theories, and also others from the social science areas to suggest alternative explanations for social and technical change, or the absence of change. These are explored in more detail in later chapters.

1.2 Major Debates about Energy Transitions

Several key themes and debates re-emerge consistently in the energy transitions literature. Many of them are about technological choices, or strategies to pursue technologies or policies to drive decarbonization. Several countries have engaged in strong efforts to transition their energy systems, with several northern European countries claiming to be fossil fuel free or 100% renewable by 2040 or earlier. In the US, efforts to decarbonize the electricity grid have led states to set targets, with some targets set at 100% clean energy. Other debates in energy transitions are less about specific technologies and policies and more about organization forms or ethical questions such as who gets to decide what these energy futures look like or who owns future energy systems (◘ Fig. 1.1).

Research into renewable energy transitions became popular in the late twentieth century around the oil shocks that affected the global transportation fuel economy. The upshot for some energy scholars is to underscore the slow pace of energy transitions, pointing to the 1940s as the last new prime movers to become dominant energy sources (Smil 2004). Amory Lovins (1976) distinguished between hard and soft paths in energy transitions. The hard path refers to coal and nuclear power, while the soft, which he advocates, is led by renewables and appropriate technologies. Lovins also suggested that distributed energy can reshape energy geopolitics, whereas centralized energy resources reinforce business as usual.

While he most typically put solar on the soft path, Lovins (1976, p. 81) warned that "not all solar technologies are soft." This focus on distributed energy resonated with Schumacher in *Small is Beautiful*, where he emphasized distributed

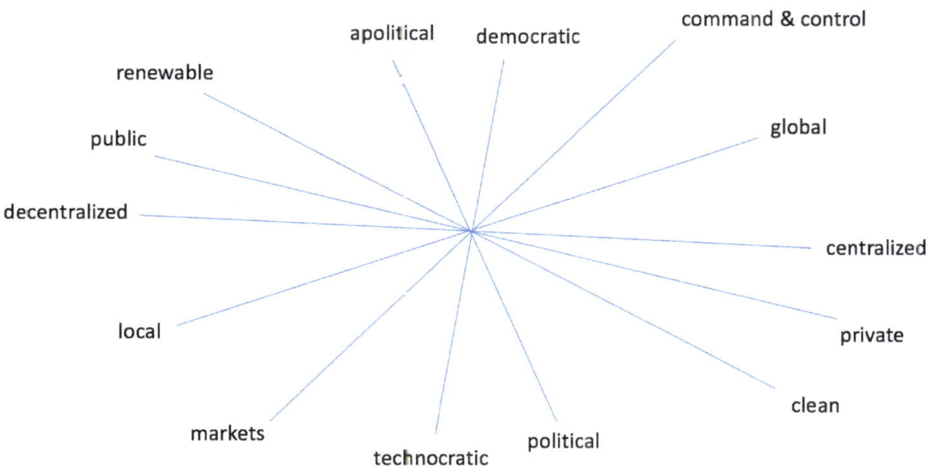

Fig. 1.1 Diagram of major axes of debates in energy transitions

energy generation as an instance of "appropriate technology"—small-scaled, decentralized, efficient, and locally controlled. Political scientist Langdon Winner argued that some forms of energy production like nuclear energy rely on authoritarian forms of social organization to protect nuclear fuel and waste (Winner 1989). Uranium in the supply chain for nuclear fuel and plutonium in the waste (or some fuels) require militarization and heavy security as nuclear power plants because of vulnerability to meltdown accidents or occasional releases of low-level radiation. Winner argues that technologies are not neutral but can have inherent politics. Many of the claims about scale and appropriate technology made in this vain are in line with claims from those seeking pathways toward degrowth. STS scholar Donna Haraway makes a similar point, but emphasizes that technologies are not neutral because they are connected to people, things, and other species, and the types of, and outcomes from, those connections matter because they are most definitely not neutral in how they change the world.

Lovins—the founder of the Rocky Mountain Institute based in Colorado, a premier renewable energy and energy efficiency research and advocacy organization—detailed these different options in 1976, in the wake of the energy crisis. Lovins' (1976) article in *Foreign Affairs*—which incidentally remains the periodical's most widely distributed article—warned against the US plans to increase its reliance on nuclear power as a path toward reducing dependence on imported oil. Not only would a large nuclear program—an order of magnitude more nuclear plants then are in operation in the US today—raise significant safety and security risks, argued Lovins, preventing the theft of weapons-grade nuclear materials would require a highly centralized and authoritarian infrastructure, with corrosive effects on American democracy and individual freedom (Ayres 1975).

Energy transitions are about more than just energy systems and economic models. There are also critical opportunities to reshape changes in household behaviors with green technology adoption, or pro-environmental behavior. It is incumbent on energy scholars to delve into questions that challenge and be reflexive about core assumptions. Chappells and Shove (2005) show clearly the importance of social constructionism perspectives and the social evolution of human expectations of things like comfort. Assumptions about occupant expectations of temperature and humidity range currently feeds into building design standards, particularly heating, ventilation, and air conditioning (HVAC) system designs. It turns out human behaviors and how user use things we make are important. But there are also debates about how well energy-saving behaviors "stick" and whether energy savings are just spent elsewhere on other energy-consuming activities—the so-called rebound effect.

Some technologies deployed in an energy transition could be accompanied by spillover benefits, especially if targeted for deployment-specific ways. Photovoltaic (PV) adoption by homeowners could have impacts on other environmental behaviors such as energy economy of vehicle purchase to outright reductions in energy use (Hondo & Baba 2010). The rebound effect described earlier—where energy efficiency adoption leads to increased energy use or energy service demand—is not well characterized for homeowners who install PVs and is an understudied topic. The integration of PV with smart grids may mark a fundamental reconfiguration of the relationship between consumers and utilities, and promises to rearticulate energy use as a social practice as consumers respond to new information and self-govern their consumption of electricity (Bulkeley et al. 2016). The term "prosumer," or producer-consumer, attempts to describe new ways in which distributed energy generation, the grid, and the built environment are integrated.

Inquiries into energy transitions—shifts in the nature or pattern of energy utilization and resources—are motivated by a host of reasons, most recently climate change. But there have also been instances where risks from price volatility, innovation policy, national security, and/or job creation were the primary motivations (Araújo 2014). Powerful cultural and political economic currents in some instances explain or constrain energy transitions (or lack thereof) as many conventional energy systems are vested with social and economic power that entrench energy pathways or block transitions, warranting an examination of the uneven geographies of energy (Juisto 2009). Incumbent energy sectors can lobby to shut down reforms. Another potential obstacle would be if the public mounted a populist revolt in response to the rising costs of energy from decarbonization efforts.

The collection of research and analysis that falls under the rubric of "social planning for energy transitions" is focused on questions about energy futures, energy policy designs, and institutions to enhance public participation, behaviors, and social acceptance (Miller & Richter 2014). There is a long history of energy, petrochemical, and infrastructure projects with very negative impacts. The general idea of this scholarship is that as energy transition will require making old industries better, and greener, newer industries and infrastructure will be required, and it will be important to have participation from the community of locals to minimize potential worker exposures, mining impacts, land use changes, and generation of waste.

Energy geographers are providing spatial and qualitative data to support decisions and evaluate impacts of alternative energy futures (Calvert 2016). Geographic research can understand the ways that a particular spatial distribution of renewable projects in, say, South Africa led to positive impacts on rural economies (Lombard & Ferreira 2015). Planning and policy must forefront the distributive impacts of energy transitions to ensure that some do not bear most of the burden from our collective response (Newell & Mulvaney 2013). This includes ensuring that even if the impacts of a renewable energy source are much less than its fossil fuel counterpart, that some people and places are not overburdened by renewable and low-carbon energy projects and impacts.

Lastly, and perhaps most importantly, is the question of indigenous and First Nations' peoples and their role and position in energy transition (Whyte 2017). This raises questions about recognition, consent, and cooperation in energy transition, particularly where resources or land, which are owned, managed, or significant to these communities, are at stake. There are also opportunities to utilize traditional ecological knowledge (TEK) in the sustainable use and management of resources. These management opportunities seem to operate best when led by tribes and First Nations, or whichever groups are most historically familiar with how to live with the land.

1.3 Degrowth Versus High-Energy Society

How much energy is needed by human civilization? Some parts of the world do not have access to modern energy, other parts of the world waste more energy than they need for the services they require. What is the population of humans in 2050 we should plan for, and how will we acquire energy for them? Will future energy demands look more like the average American consumer in the early twenty-first century, or will it more closely resemble energy use in Europe? There is some minimum threshold of energy to ensure that human well-being and quality of life needs are met.

The 2,000-watt society is one vision put forth by the Swiss to describe a target for a lower energy use society (Notter et al. 2013). 2,000 watts of power is about 48 kilowatt-hours (kWh) of energy per day, and the Swiss argue this target is balancing basic human needs against overconsumption of energy. One thing is for certain, if there is less energy consumed on aggregate, it makes meeting climate and air pollution goals much easier and also lowers impacts from the energy sources that unseat fossil fuels.

Trends suggest that energy demand globally is still increasing, however. Data available from International Energy Agency's *World Energy Outlook* (cf. IEA 2016) annual reports and BP's *Statistical Review of World Energy* (2019) confirm this point. 1990 is often considered a milepost year in climate benchmarking because the first major global climate policy of the Kyoto Protocol used this date to measure emissions reductions. The world population was 5.3 billion in 1990, and annual electricity consumption per capita was 2.07 MWh per person. By 2015, the world population increased to 7.3 billion and electricity consumption per capita was 3.05 megawatt-hours (MWh) per person, roughly 50% more per capita, with overall electricity demand around 18,000 terawatt-hours (TWhs) (IEA 2016). Human civilization could depend on increasing amounts of energy in the future.

In parts of underdeveloped countries like China and India, and across regions of sub-Saharan Africa, the Middle East, and southeast Asia, access to energy services is an urgent matter of increased living standards. China comprises 19% of the world's total population. Such a large population and industrial production requires an immense amount of energy. In 2003, 31% of the world's coal was consumed by China, eventually increasing to 50% by 2013 (Bloch et al. 2015). As a result of burning so much coal among other things, China has the highest rates of lung cancer in the world. From about 2008 to 2013, China was building a new coal-fired power plant every week. Yet, even this trend seems to have receded. Not only has coal-fired generation slowed, China added more solar than the US ever installed for each of three consecutive years 2015–2017, and was already the world's leading generator of wind power. One energy future scenario anticipated reduced overall global energy demand by 40% in 2050, about 245 exojoules (6.8×10^{13} or 68 trillion kWh), even with population growth, economic activity, and increased well-being (Grubler et al. 2018).

While these trends toward a higher-energy society continue to march forward, there are some that contend these trends should be reversed. Building on many decades now of work in ecological economics and political ecology, Giorgos Kallis (2011) is a leading writer and thinker on the question of degrowth. The basic idea underlying degrowth is that human civilization needs to discontinue its ties to an economy where the growth imperative drives much of the economic decision-making and underlies measures of economic and social well-being. The degrowth movement aims to redesign economies so that they are more compatible with a reality that is subject to the laws of entropy. The fundamentals of degrowth may seem heretical to mainstream economics, namely, the idea that growth, measured by increased gross domestic products (GDPs), is the best measure of the health of the economy. Instead, degrowth thinkers look for examples of energy systems and economies that are steady-state, that are consistent with the avoidance of waste implied by the entropy laws, and that exhibit more democratic, anti-racist, and decolonial tendencies. Degrowth in its contemporary form rejects population-based arguments of neo-Malthusianism, though older, less careful writing suggested more connections between the philosophies. The basic ideas underlying degrowth were first developed by Nicholas Georgescu-Roegen, a Romanian ecological economist, and popularized by his contemporaries Herman Daly, Kallis, and others. Entropy laws mean that conversions of energy result in less useful work available with each conversion. This is why the laws of thermodynamics dictate that the end of the universe will be a cosmic heat death. The total amount of energy will be the same as it always has, but none of it will be available to do work. That means no light, no movement, and no heat.

A key principle of ecological economics is to avoid production systems that convert low-entropy resources inefficiently into high-entropy waste. This often means harnessing energy resources that rely on renewable flows: sunlight, wind, plant growing cycles, or heat from below ground. This is important because instead of cycles of capital investments producing widgets and circulating human, financial, natural, and manufactured capital, entropy laws suggest production is more linear in nature as natural resources are extracted, degrading the quality of energy available, produced, distributed, and then disposed of as waste. Since natural capital and low-entropy resources are the limiting factors of economic development, any effort to incorporate sustainability needs to break these flows and instead make

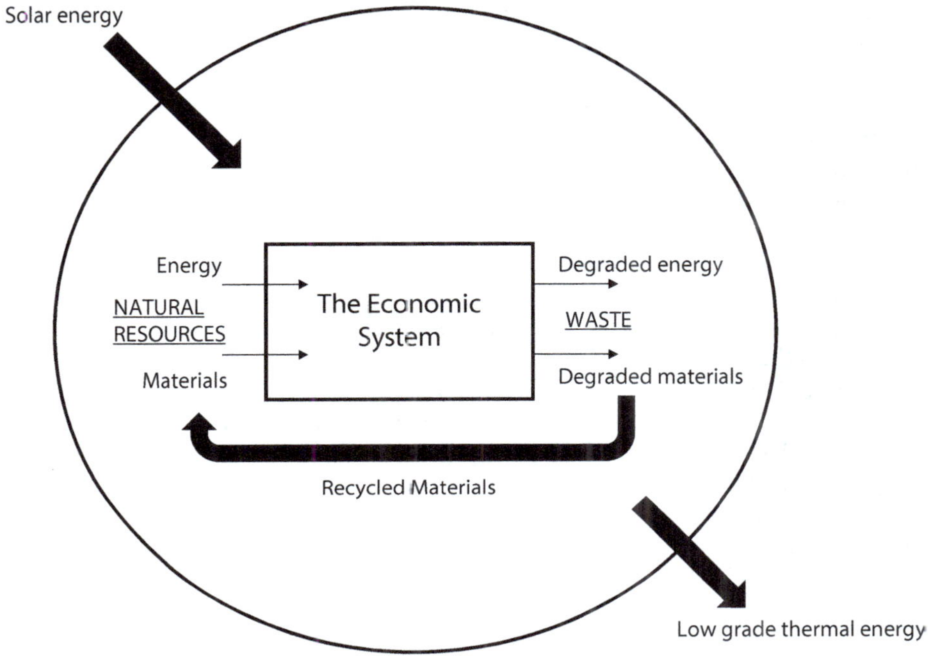

Fig. 1.2 Entropy and ecological economics

a truly circular economy. This means the outputs from some economic activities are recycled as inputs to other activities (Fig. 1.2).

1.4 Low-Carbon Resources: Clean Energy Versus Renewable Energy

Energy transition debates at some point will encounter questions about whether nuclear power and/or natural gas should be considered as sustainable energy resources or whether a renewables-only approach is warranted. A system based on renewables could be produced a number of different ways, including across great distances, with more energy storage, and by pursuing extensive demand response. There are good arguments on both sides that argue that their take is the most important. The reality is that there are a number of different ways low-carbon energy resources can be developed, deployed, and integrated, and the degree to which they include nuclear or natural gas may be geographically dependent.

Some natural gas power plants can be operated in a flexible manner, whereas most nuclear power plants (older plants in general) cannot be ramped up and down. Others find that nuclear power has opportunities to ramp up, and has already used this method in a number of electricity grids (Jenkins 2018). Although some nuclear power plants could be reconfigured to other tasks such as desalinating water or pumping water for pumped storage, another method to displace greenhouse gas (GHG) emissions is taking waste heat from natural gas or nuclear power plants and using it for district hot water heating.

Around 2007, when the Nobel Peace Prize was awarded to the IPCC, there was a nuclear renaissance brewing. The US for the first time was planning to develop new nuclear reactors, and there was great interest in making nuclear power a key tool in the response to climate change. Two reactor technologies, light water reactors (LWR) and boiling water reactors (BWRs), are the primary nuclear plants operating in the world. But the US had not ordered a new nuclear power plant since 1978. Some attribute this stoppage to the accident at Three Mile Island. But the reality is that huge cost overruns, lengthy delays, flattening demand for electric power, and expensive maintenance led to quite difficult economic challenges to the nuclear power industry in the US.

The 2011 accident at the Fukushima Daiichi nuclear power station in Japan changed all of this. Germany already was planning to abandon its nuclear power plants, but the earthquake and tsunami had the immediate result of taking nearly all nuclear power plants in Germany and Japan offline, and hastening plans for their early retirement. Germany has taken quite a bit of criticism for closing nuclear power plants, while coal-fired generation still operates there (Steinbacher & Röhrkasten 2019). Japan replaced much of its lost power generation with liquified natural gas (LNG).

Many advocates of nuclear power and energy analysts embrace the idea that modular reactors could be cheaper and safer to operate than those still in use today because they would be manufactured in mass production facilities and can be designed with passive cooling (cooling the reactor core does not require external energy inputs in these schemes). Some proposed future nuclear power plant designs include combined heat and power or desalination applications in by design order to utilize all energy, including waste heat, and to make plants operate more effectively. Old challenges like nuclear weapons proliferation and high-level radioactive waste disposal, and hopes to use nuclear to mitigate GHG emissions, produce next-generation passively safe reactors, and to turn nuclear warheads into electricity—swords to plowshares—will shape continued debates on nuclear power.

A somewhat cornucopian vision for natural resources, eco-pragmatists assume high-energy society is here to stay, directly challenging to the notions of scarcity underlying the *Limits to Growth* (Meadows et al. 1972). In the *Eco-Modernist Manifesto*, they posit that "a new generation of nuclear technologies that are safer and cheaper will likely be necessary for nuclear energy to meet its full potential as a critical climate mitigation technology" (p. 23). They point to France and the long history of relatively safe operation of the US nuclear fleet, which has exceeded expectations in terms of plant age and capacity factor (percentage of time the plant operates). They argue that nuclear power is the only low-carbon energy technology capable of fully meeting the urgent response needed to climate change.

Advocates for incorporating natural gas into energy transition strategies point to it as the lowest carbon fossil fuel and most efficient to combust. The lowest GHG emissions factor associated with fossil fuel combustion without carbon capture is a combined cycle natural gas (CCNG) turbine. When CCNG plants displace old, inefficient coal-fired generation, there is a considerably lower amount of GHG emissions per unit energy. Methane emissions from natural gas infrastructure and fracking erode some of the GHG benefits. But perhaps blending natural gas with biogas or capturing carbon emissions will offer opportunities to lower carbon emissions.

Can we get to a low-carbon-emissions scenario with renewables—wind, wave, solar, hydro, geothermal, and biofuels—alone? Will the falling costs of renewables continue such that solar and wind energies will be the lowest cost energy sources? How will the problem of intermittency with renewables be solved?

Is it possible to power human homes and infrastructures with 100% renewable energy pathways? One version popularized in a *Scientific American* article in 2009 is the wind, water, solar (WWS) strategy (Jacobson & Delucchi 2009). The plan requires that all end uses—heating, transportation, and so on—utilize electricity, the primary energy carrier produced by various renewable energy sources, including wind, hydroelectric, and concentrated solar and photovoltaic power. The plan relies also on lesser amounts of geothermal, wave, and tidal power, but it does not include any combustion, excluding biofuels often described as renewable fuels. The WWS includes electrical, heat and cold energy storage and increased hydropower capacity to help account for the intermittency with renewables problem. Jacobson's work on air quality and combustion led their team to eschew biofuel technologies, showing it is not possible to meet air quality standards burning ethanol because high levels of NO_x are produced with combustion of gasoline-ethanol blends; NO_x is produced in all combustion since more than three-quarters of the air is comprised of nitrogen. The authors produced a follow-up analysis for the journal *Energy Policy* (Jacobson & Delucchi 2011; Delucchi & Jacobson 2011) and the *Proceedings of the National Academy of Sciences* (Jacobson et al. 2017), and they have produced numerous versions of WWS strategies for various countries and regions of the globe.

The first WWS paper proposed renewable energy deployment on a worldwide scale and seven ways to architect and implement a worldwide renewable energy system so that it will reliably satisfy demand and not have a large amount of capacity that is rarely used:

> (1) interconnect geographically dispersed naturally variable energy sources (e.g., wind, solar, wave, and tidal)…(2) use a non-variable energy source, such as hydroelectric power, to fill temporary gaps between demand and wind or solar generation…(3) use 'smart' demand-response management to shift flexible loads to better match the availability of WWS power…(4) store electric power, at the site of generation, for later use…(5) over-size WWS peak generation capacity to minimize the times when available WWS power is less than demand and to provide spare power to produce hydrogen for flexible transportation and heat uses…(6) store electric power in electric-vehicle batteries…(7) forecast the weather to plan for energy supply needs better (Jacobson et al. 2017).

The WWS project became popular with the climate activist community with the dissemination of reports through the group, The Solutions Project. The group first produced 50 state solutions for the US and later for 139 countries (Solutions Project 2018). *Gasland* Director Josh Fox, activists, and Hollywood personalities—Greta Thunberg, Mark Ruffalo, and Leonardo DiCaprio—have taken up the 100%

renewable energy mantle by becoming outspoken public advocates for climate action. Several US Senators and state governments have proposed standards for 100% renewable electricity, and the California legislature passed a 100% clean energy law in 2019 that essentially functions as a 100% renewables goal. This Hollywood attention and support from other organizations and politicians has hoisted this plan to electrify everything with renewables to be a central pillar of the progressive left with Congress woman Pramila Jayapal, even Senators Jeff Merkley, Corey Booker, Ed Markey, and Bernie Sanders sponsoring legislation for 100% renewable electricity (Merkley 2017).

There are some obvious challenges to attaining 100% renewable energy, with the most challenging being the integration of variable intermittent resources to ensure a reliable grid. These are not necessarily technical issues, but rather economic ones. Ensuring that there is enough power throughout the year, across the seasons, and even dealing with year-to-year variability could result in a significant overbuild of renewable resources, by adding storage, or reducing demand. One proposed solution requires installing several times over the power capacity of renewable power plants to compensate for intermittency (Loftus et al. 2014). Solutions to the challenge of intermittency suggest that on-demand energy generators or storage devices must be available when the wind stops blowing and after sunset. Given the difference in the availability of renewables and demand across seasons, there may need to be seasonal storage. Energy storage, heat sinks, demand response technologies, and load-following generators like hydropower are critical to making the WWS strategy work. Additionally, land-use change becomes an important consideration in the shift from subterranean fossil fuel resources to energy collected on the surface of the Earth (Smil 1984).

There are reasonable critiques of the "electrify everything" approach. Electricity is the highest quality energy carrier, when many energy applications are only for low-quality heat. Electrifying everything may be overdoing it, producing higher quality energy than is needed. The WWS approach to 100% renewables also overlooks opportunities to acquire renewable energy from waste resources (e.g., some landfill, dairy, or waste treatment biogas). But if renewables are cheap and abundant enough and can be deployed with minimal life-cycle impacts, overbuilding renewables may not be a bad thing. More than 20 authors with expertise in energy systems published a paper in the *Proceedings of the National Academy of Sciences* criticizing the WWS strategies (Clack et al. 2017). The key contention is that the model underlying the WWS strategies failed to appropriately account for the real-time ramping up needed to correct intermittency issues from the 100% renewables load. One problem identified is the lack of hydropower generation or power system modeling. WWS models assume far more capacity than needed as they use hydropower as the load-following resource and during long periods without sun or wind. The critics contend that the necessary hydro capacity needed in WWS models is an order of magnitude higher than the total installed hydropower capacity in the US, adding that it could flood downstream communities by pushing very high flows over short durations. Additional turbines could be added, or batteries or other storage technologies or technical proxies could be used (demand response, ice storage/chilled water for cooling). Other contentions raised in the critique include questions about how WSS strategies conclude reliability without modeling loss of load probability or frequency regulation (making sure grid hums along at 60 hertz). The WSS strategies also do not appear to include any transmission model and assumes that about two-thirds of the power demand is from flexible sources.

There is a camp in the energy space that looks at 100% solar energy pathways, or ones where the technology provides the overwhelming majority of power. The long view on energy supplies from available sources points to a future that requires a transition to solar energy, just because it is so widely available, even if it is dispersed. This point was not lost on Thomas Edison, who apparently said in 1890 something to the effect of wagering a bet on the sun as the energy resource for the future.

Solar energy technology researchers proposed a "solar grand plan" for the American Southwest (Zweibel et al. 2008). The idea advocated a technological mix based on electricity generated from cadmium telluride (CdTe) photovoltaics, wind, and concentrated solar power with molten salt storage to be deployed in the American Southwest with various storage technologies and complementary renewables, with an emphasis on compressed air connected to wind farms. The solar grand plan proposal aimed to create large solar power plants that can generate approximately 69% of the electricity needs of the US as well as 35% of its total energy by 2050. Because there are currently still nuclear and coal power plants that are in operation, retiring these inflexible generating plants will help reduce GHG emissions significantly as renewable sources of energy are becoming much more available. Currently, there are four key technologies that could be implemented: photovoltaics (PV), compressed-air energy storage, concentrated solar power, and high-voltage direct current (HVDC) transmission.

Energy storage will remain a big challenge. As of 2020, the majority of global energy storage is in the form of pumped hydroelectric storage, where energy is used to pump water against gravity to an elevated location, and then when energy is needed, the water is released to spin a turbine and generate electricity. But other storage technologies, most notably lithium ion batteries, are rapidly growing. Integrating storage into the electrical grid could help a state like California avoid curtailment—stopping the delivery of electricity—from intermittent wind or solar power (California ISO 2017, p. 2).

1.5 Distributed Versus Centralized Energy Systems

A major axis in debates about energy transitions is the extent to which energy systems need to be distributed or centralized generation. Modern electricity grids are centralized, with large power plants pushing electricity along high-voltage transmission lines, dropping the voltage down to distribution, and delivering to homes and businesses. But some envision a different electricity system, one where there are small power plants scattered throughout, delivering power along distribution circuits that move power in both directions, possibly allowing customers to disconnect or operate autonomously from a larger electricity grid. This is a cornerstone debate in energy transitions, and it is relevant to contemporary debates that are circulating in energy policy discussions even today. When a hurricane named Maria hit Puerto Rico in 2017, it revealed a tremendous amount of vulnerability of highly centralized electric systems. Similarly, Hurricane Sandy, which hit the northeast US in 2012, led to major long-term blackouts, despite many homes with rooftop photovoltaics in states like New Jersey, because the only way the system worked was by its connection through a large grid. On the other hand, we might see a system of interconnected energy systems that can operate independently with much more resilience. But that comes with its own challenges.

Can enough power be generated locally for reliable power? How far do interventions like energy efficiency, demand-response, distributed renewables, smart grid innovations, and changes in social practice take us toward decarbonization? What is the future organization model, and also business model of the electric utility? What is the balance of behind the meter versus electricity in other grid domains, or even off-the-grid? How far afield will networks of electric power draw from for resources? These are all critical questions that will require careful planning and deliberation to resolve.

These sentiments were emphasized by science and technology scholar Langdon Winner, who argues that technological artifacts such as nuclear reactors have important implications for social order because of their entanglements with questions of national security, geopolitics, and nuclear weapons proliferation (Winner 1986). Socio-technical systems, such as those that make possible electrification of human society, have tendencies and reverberations independent of human intention. Different technological paths have different socio-ecological arrangements and impacts and, by extension, effects on human well-being and security. In other words, technologies are inherently political, comprised of multiple nature-society associations, and sometimes have unintended effects.

The appropriate technology folks also argued that, by contrast, solar energy technologies did not rely on either risky or faraway energy sources and were more amenable to decentralized modes of energy procurement and generation. Obviously they did not account for the mining and extractives needed for renewables, but these features, they claimed, would promote energy autonomy, democratization, and even reduce the political and economic power of electric utilities and energy corporations. Solar energy is freely available to anyone able to capture it *on-site*, making it suitable for decentralized, small-scale, distributed power generation (DG). PV technologies, in particular, enable people to electrify homes and businesses with limited reliance on or even no connection to the electricity distribution grid, fostering a sense of energy autonomy and freedom, and enhancing well-being and security. Already in the early 1970s, *The Whole Earth Catalogue*, the counterculture was thriving and one could order photovoltaic systems. Stuart Brand's counterculture publication offering "designs for living" included plans and advertisements promising PV systems and Buckminster Fuller geodesic domes that could increase the energy self-reliance of communities and individuals, and the country as a whole (Turner 2006).

A book from the 1990s, *Who Owns the Sun?* attempts to get at the possibility for a society where energy resources are distributed and decentralized, linking it to notions of democratization (Berman & O'Connor 1996). Along similar lines, E.F. Schumaker's widely read book *Small is Beautiful* would be a mainstay of the appropriate technology movement. The key attributes of appropriate technology are that they are small-scale, decentralized, labor-intensive, energy-efficient, environmentally sound, and locally autonomous. They are people-oriented technologies. Schumacher's "small is beautiful" refers to decentralized solar power, rainwater harvesting systems, biogas, and passive solar home design, much like the Earthships of Taos, New Mexico, or Drop City, Colorado (replete with their sustainable mobility contradictions). Lovins argued that energy supplies should be compatible with intended uses and showed how residential homes could supply enough power for a home. Technologies need to be small, low impact, and decentralized to be green according to this logic. Key ideas underlying the appropriate technology movement can be traced back to Mahatma Gandhi, who looked favorably on technologies that were at the scale of the user, perhaps earlier.

Applying the small is beautiful concept to energy transitions posits an energy future led by concepts like smart grids, microgrids, energy watersheds, demand response technologies, distributed fuel cells, photovoltaic modules, and heat pumps, essentially all the conversion devices that work in decentralized power systems There is a strong emphasis on flexible and resilient loads, and energy efficiency is a core principle in system design. The emphasis on flexibility is a reflection of the intermittency of renewables and the need to have the capacity to draw on or store power when the demand and supply fluctuates.

Global electricity systems cross most continents on Earth. The biggest electricity operators in the world are in the US (the grid that provides electricity to a region extending mainly across Pennsylvania to New Jersey to Maryland), and in France. Many electricity grid operators function within three major interconnections in the US—the Eastern, which is the Midwest to the eastern coast; the Western Electricity Coordination Council (WECC), which includes Canada and Mexico; and most of Texas operate their own electricity grids. Layered on top of this are balancing authorities—that are there to plan for the movement of energy to ensure the grid operates smoothly; 38 balancing authorities operate in the US as of 2018, and there are many more globally. These balancing authorities are organizations that work with electric utilities and independent merchant power generators to ensure that electricity is available across the region they are responsible for, and they help coordinate electricity sales between these nodes in the system. China operates across mostly five grids run by the State Grid Corporation of China, the biggest single power company in the world. The entire continent of Africa delivers electricity to some places via five power pool systems, each tied to the urban areas of Africa such as around South Africa, Northern Africa, the Nile Valley to the horn of Africa, Lagos, and places to the north and west. Much of the continent remains without access to electricity. Google "electric grid" and the place in the world you are interested in next time you have a chance.

We are going to need to build a lot of clean energy infrastructure. Forms of ownership and control can influence how organizations respond to the need to transition. Municipally or community-owned power plants can be more easily accepted by the public and less likely to face social resistance. Privately owned companies may have an advantage where there is a need to draw on more capital for investing in new equipment. On the other hand, public or cooperatively run facilities or organizations will have more impetus to reinvest all revenues into the equipment, whereas private companies may direct some portion of revenue to profit or dividends on stock. Some municipalities may even be able to borrow cheaply on the bond market.

In areas where there is increasing amounts of solar power being delivered to grids, the duck curve is a widely used visual metaphor in the energy policy and electricity system design community. The duck curve refers to a graph of effective load, or demand, of electrical energy, and it has a distinctive dip, resembling the belly of a duck, during midday hours when solar power resources are operating at or near full capacity, reducing the need for fossil fuel generators. The neck of the duck is the part of the graph where demand increases rapidly as the sun sets. The general idea is to flatten this curve by pushing some of the solar power generated at midday to the evening to reduce the evening ramp up of fossil fuel generators via energy storage or peak displacement.

Microgrids are another electricity infrastructure strategy. They can have many elements of traditional electrical grid, but they are smaller in scale and use information to help manage the system to balance demand and supply. Microgrids tend to have energy storage, which makes them more adept to handle intermittent energy sources such as wind and solar. Microgrids have increased resilience as a result of decentralization, and also advanced communication technologies can help utilize information about demand and even weather patterns to anticipate operational strategies. Microgrids can function independently even when the connected grid goes down. They may help to serve remote communities that are too difficult to connect to a larger main grid, by being an entirely independent and lightweight remote solution. In California, they may be a solution to catastrophic wildfires caused by failing electricity transmission and distribution equipment. Thanks to these characteristics, microgrids are becoming an increasingly important new technology in island communities, remote regions, and in disaster relief centers.

Finally, there is quite a loud buzz around blockchain technologies like Bitcoin to help facilitate payments and exchanges in a decentralized, distributed network of energy buyers and sellers. The idea is that there are too many transaction costs related to the buying and selling of small amounts of energy, which might be very critical to a smooth functioning grid running on intermittent resources. These distributed currencies (as opposed to centralized currencies like the US dollar, Indian rupee, etc.) operate with a ledger that tracks all transactions, making these verifiable and low-cost transactions. However, there were numerous reports suggesting that Bitcoin, the most widely known of these currencies, required copious quantities of energy, more than entire nations in some speculation (Truby 2019). This is because "mining" for Bitcoin requires copious amounts of computing power.

1.6 Deployment Versus Breakthrough Technologies?

How far will we get by deploying existing technologies? Or, are breakthrough technologies needed to get us from here to there? This is another axis in conversations about energy transitions. Some energy experts, like Stephen Pacala at Princeton University, Dr. Katharine Wilkinson from Project Drawdown, and Carl Pope at the Sierra Club, have argued for many years that we have technologies today to make significant inroads in displacing greenhouse gases, and we should deploy.

Other groups and organizations emphasize the need for research and development. Think tanks like the Breakthrough Institute and personalities like Bjørn Lomborg argue that technologies need to be developed and made more efficient before deployment. In the real world, both occur simultaneously.

What innovations are best learned through deployment? Many researchers argue that supply chain development and learning by doing in the manufacturing process are where innovations in cost savings are most likely to occur. The incremental benefits in technological shifts, which occurred on an annual basis, could lead to critical innovations. Scaling-up photovoltaics manufacturing and supply chains over the past decade has led to much higher yields on the manufacturing line, more powerful panels, and reduced costs through economies of scale.

Energiewende in Germany

In Germany, energy transition is fueled by policies derived from the Energiewende plan, literally translated as "energy transformation." A German energy transition was initially led by large-scale wind farms and soon after large-scale photovoltaic manufacturing. These also came at a political cost, what many dubbed "the latte fallacy"—the idea that the extra cost of energy would be about the cost of a latte coffee per day. These low costs never materialized and instead the energy system cross-subsidized some of these customers, creating some political backlash. The challenge with first movers along this pathway is that early technologies are going to cost more. Prices will come down as technologies scale up. That was the goal of the California rooftop solar initiative that subsidized the earliest adopters the most, a subsidy that stepped down concomitant with the adoption queue. German's feed-in-tariff led to more expensive than anticipated electricity because prices did not fall until much of their market was saturated. Germany's early solar push helped drive down the cost of solar energy through deployment, and these benefits were reaped by later adopters of technologies who were able to buy-in at lower prices.

1.7 Natural Capitalism or Ecological Socialism?

The type of social arrangements in the economy are sometimes the core contentions in energy transition debates. There are some social theorists who contend that capitalism is not compatible with sustainability because it is not possible to have endless growth on a finite planet (sounds like degrowth, right?). Some go as far as to say collectivization and elimination of the profit motive will be required to move civilization onto a path towards sustainability. They embrace ideas like the sharing economy, collective ownership and coops, guaranteed minimum incomes, and communing. In practice, there are attributes of both incorporated into different systems and institutions for energy transitions. There are services that offer people an opportunity to share cars and bikes today, but within the private sector model. This hybrid model takes elements of one (sharing) and the other (profit to private companies).

Economists describe capital as coming in human, financial, and manufactured capital forms. But if the Earth is what provides the natural resource base, and this is not infinite, it is also important to consider natural capital. Arguably, natural capital is the limiting factor in economic development. Paul Hawken, Amory Lovins, and Hunter Lovins show how our wasteful economic system could be remade by adhering to a handful of principles to incentivize sustainable practices (Hawken et al. 2013). The first is radical resource productivity, essentially being more energy and material efficient in manufacturing and natural resource use. Second, biomimicry is a principle that points to opportunities to learn from nature and incorporate those lessons into design. Third, transitioning to a service and flow economy from a goods and services economy is thought to save on redundancy and take advantage of opportunities to share.

Investing in natural capital is the final key principle embraced here as the idea is that natural capital produces more natural capital and, therefore, contributes overall to that which can be valued. This is what it means when sustainability practitioners talk about abundance. The basic way to get to a natural capital economy is to internalize externalities in markets or embed them in prices. Ronald Coase introduced ideas about how to deal with market failures and internalize the full costs in the 1940s: put the social costs back into the prices.

Policies conformed to fit capitalism's affinity for market solutions are not without failings. In the 1970s, sociologist James O'Connor suggested a second contradiction of capital. The idea is that capitalism tends to erode its own resource base. This is a similar point to early thinkers like Lewis Mumford and others. So for many eco-socialists, natural capitalism suffers this fatal flaw of not being capable of a response to undermining its own resource base. Somewhere in between are those focused on questions of energy democracy, today, where questions about who owns and controls energy are core.

1.8 Socio-Technical Systems and Multi-Level Perspectives

How long do energy transitions take? What is the scale, duration, and the magnitude of energy transitions? These are two poles of important debates in energy transitions about scale and pace. On the one hand, scholars like Vaclav Smil argue that energy transitions are glacially slow. Smil uses the concepts of prime movers and energy converters to describe the kinds of technologies and points out how slow each transition took to move from one to another technology or source. There are of course some sectors where conversions can happen more swiftly, but for entire economies, the pace is much slower according to this view. Some researchers argue that there are examples of very deliberate energy transitions that occurred much more quickly in the past. In 2019, the UK had many consecutive days without coal electricity, and an even longer run in 2020. Coal was the dominant primary energy used in the UK for electricity just 50 years ago.

What are the drivers of energy transitions? Is it about getting energy prices and externalities right? How important are other policy tools? Do we also have to hope habits, cultures, and behaviors change too? Some researchers suggest that society gets stuck in technological pathways, which lock-in some technologies and energy sources because there is so much infrastructure already committed and economies of scale at work. That said, changing our technologies and resources might be easier than changing how we act.

Definition

The socio-technical systems approach to understanding social and technological change emphasizes the interactions and new social orders that give rise to new relationships between humans, each other, and their technical devices. Energy transition scholar Frank Geels describes how new "system innovations not only involve new technological artefacts, but also new markets, user practices, regulations, infrastructures, and cultural meanings" (Geels 2004).

Some socio-technical system researchers propose the multi-level perspective as a framework to explain change by describing interactions between nested elements of—niche, regime, and landscape. The basic idea is that as low-carbon technologies become available, they first begin to substitute for other energy sources, which leads to transformation, reconfiguration, and de/realignment across these nested areas.

The political scientists who study energy transitions assume that some kind of external shock is required to facilitate energy transitions (Akin & Urpelainen 2018). They argue that alternatives are ushered in as a response to focusing on events: deadly air pollution inversions, catastrophic oil spills, political dynamics underlying the oil shocks of 1973 and 1978, or other geopolitical or strategic motivations. These theories of social change underly narratives that emphasize the worst-case scenarios of climate change in policy decision-making, where the idea of having a focusing-event, like a natural disaster enhanced by climate pollution, is believed to motivate climate policy.

1.9 Supply-Side Strategies: Keep it in the Ground, Divestment

Most of the energy strategies discussed in the transitions literature focus on shifting resources and technologies from existing generation technologies to renewable or clean energy technologies. There is another camp of transitions advocacy that focuses on limiting the supply of fossil fuels. They argue that instead of the consumption-based emissions reduction pledges, nations need to commit to lowering fossil fuel production. This supply-side approach aims to "keep it in the ground" by identifying the potential carbon pools likely to be burned by energy companies. The organized foundation for this movement in recent years is the article popularized in

Rolling Stone by Bill McKibben (2012). The article shows how much carbon that energy companies carry on their ledgers as value for shareholders, which they estimated to be 2,795 gigatons (Gt) of carbon from coal, oil, and natural gas reserves if these fossil fuel reserves were burned. Climate scientists suggest that only 20% could be safely burned (565 Gt) without exceeding important climate thresholds. This extra fossil fuel resource base—the 80% that must stay in the ground—is commonly referred to as unburnable carbon. The problem is that the amount of carbon valued by energy firms is five times more than the capacity of Earth's atmosphere to absorb those emissions without exceeding 1.5 or 2 °C of warming. The analysis points to the need for energy transitions to also address the supply of fossil fuels. Many oil and gas company valuations are based on holding assets that are considered by climate scientists to be unburnable carbon or potential stranded assets, leading to some speculating about a carbon bubble (Lucas 2016).

With specific targets for removing some of the unburnable carbon assets held by energy companies, this allowed shareholder groups, public lands advocates, and other groups to begin targeting specific fossil reserves held by companies or countries. The Yasuní-ITT Initiative in Ecuador was an unsuccessful organized effort to pay the country to keep one billion barrels of oil away from development (Sovacool & Scarpaci 2016). Similar efforts in the US have focused on protecting federal lands from fossil fuel development (Center for Biological Diversity 2015). One critical question is whether supply-side approaches restricting oil and gas production in one place simply increases production elsewhere. If there are economic effects, such as increased costs associated with that second area, does this hasten energy transition?

This movement also asserts that there should be no more investments in fossil fuel infrastructure, as this just delays energy transitions to clean and/or renewable energy sources. This is a key assertion regarding oil and gas pipelines but also natural gas power plants. This has become more formalized as a divestment movement seeking to rid institutional investors, organizations, and estates of any stocks/shares in fossil fuel companies.

Universities are leading the way with divestment efforts, with hundreds of universities already pledging to divest and nearly 600 campaigns organized at universities around the world (Grady-Benson & Sarathy 2015). The mostly student-led campaigns have been effective at shifting divestment questions toward university sustainability programs, alongside the American College and University Presidents' Climate Commitment (ACUPCC) and the Association for the Advancement of Sustainability in Higher Education's Sustainability Tracking and Rating System, or STARS. The premise is that if universities represent an important part of visioning the future for human civilization, they should invest in energy systems of the future as well.

All strategies must overcome climate denialism and cheap energy populism, two major obstacles to public support for decarbonization and sustainable energy strategies. Climate denialism is the assertion that the science is unsettled and not known well enough to act on climate forecasts or switch away from certain types of energy supplies (Wolf & Moser 2011). Here, the solution is overcoming basic scientific literacy in the general public and also understanding the motivations for the way people object to the policy movement to respond to concerns about climate change.

Cheap energy populism is the idea that Americans have a requirement for cheap energy, and that as prices rise, there becomes a populist backlash looking to lower the price of oil. Chants of "drill baby, drill!" loom behind every uptick in fuel prices, energy companies are also not trusted by the public, and misinformation about renewables informs some of the public support. Understanding these elements make it all the more important to decenter climate in the politics of energy and look for synergies between climate and other goals, such as public health, security, reliability, and long-term savings. These are critical points to address because energy transitions that do not directly respond to concerns about rising prices risk counter-mobilization by social movement actors who could enlist populist support for keeping prices of energy low.

1.10 Demand-Side Strategies: Changing Behavior and Social Norms

Amory Lovins famously quipped that people do not really care for the type or source of energy they have. They just want warm showers and cold beer. Figuring how to provide warm showers and cold beer some other way is what some energy social scientists and policymakers call demand-side strategies. These strategies often entail swapping grid for renewable energy or energy efficiency improvement for end uses. But more broadly, demand-side strategies refer to the behavioral aspects of energy, technology choice, social norms, lifestyles, and the realization that increased energy (beyond some minimum needed for basic needs) does not necessarily result in better well-being. Researchers engaged in this area include psychologists, geographers, sociologists, economists, and interdisciplinary researchers, each with their own set of questions and methods. Economists refer to this as derived demand. This means that there is an intermediary product between the ones demanded and what it takes to deliver it. To riff off Lovins, people just want their electrical outlets to provide power and vehicles to move. In some ways, this makes energy transition easier from the demand perspective because there are no cultural reasons that someone or people want to stick with a particular source of electricity as long as it is reliably consistent. There are deep cultural ties to natural resource extractive side, so it certainly matters where electricity comes from But most consumers take this for granted.

Important topics that demand-side strategies must deal with include questions around how people get around. The car culture of the Western civilization has become very deeply entrenched, making it seem less likely that biking and walking are low-carbon solutions without very significant behavioral changes or adjustments to infrastructure that make walking and biking more safe and convenient. Perhaps the social distancing from COVID-19 will begin to change some of this. That said, much of the industrialized world already has deep bicycling and walking culture; if those modes of transportation can be illustrated to maintain well-being and strong economies, and help upward mobility, there are tremendous opportunities and potential GHG avoidance associated with pursuing these kinds of mobility strategies, including the possibility of mode shifting away from personal automobiles toward other forms of mobility, including telecommuting and enhanced virtual reality.

More difficult habits, behaviors, and routines to change include eating and food. Cultures of food are much more deeply inscribed and more difficult to change and move. Livestock direct GHG emission from manure management and enteric fermentation are a major source of emissions, rivaling the emissions from the electricity sector. Adding in emissions associated with land-use change and livestock are major contributors to climate change. Emerging protein sources have attracted significant influxes of venture capital in Silicon Valley, suggesting this could be an area of growth in the near future.

Other areas of interest in this space include understanding the needs and expectations of cultures of comfort. About 10% of greenhouse gas emissions are associated with space heating and cooling. To what extent are humans capable of living inside buildings and homes that allow for warmer and cooler indoor temperatures? In other words, to what degree is comfort a social construction? To what degree can indoor temperatures accommodate low-carbon solutions that lower heating thermostats and raise air conditioning thermostats? This is a critical question as countries experiencing extreme heat with high populations will demand air conditioning to improve well-being as their economies mature and purchasing power increases. To what degree has a chilled room in summertime become an expectation, and to what degree is it needed to improve peoples' quality of life? An assessment that focuses on human needs might help approach this topic in a way that acknowledges the need for social equity in energy transitions.

1.11 Just Transitions

Much of the spirit of this exploration of socio-ecological systems will invoke a just transition. Scholarship on just transitions thinks about the applications of ideas and principles of environmental justice to decarbonization policies. Who wins and who loses in energy transitions? The term energy justice describes the goal of equitable outcomes from energy extraction, production, delivery, and fate (Jenkins 2018). But since the term "just transition" encompasses more than just energy, it is worth differentiating these terms. Surely, energy transition will be accompanied by agricultural, water provisioning and use, and forestry and land-use transitions as well.

One helpful concept to evaluate energy justice is the concept of embodied energy injustice (Healy et al. 2019). The idea is that technologies require inputs from suppliers, and ultimate natural resources form extractive industries. If these supply chains harbor environmental injustice—say, for example, at a rare earth metal processor—this becomes an embodied injustice carried along the supply chain. So when people refer to "blood batteries" and "ethical batteries," they are referring to addressing these embodied energy injustice (◘ Fig. 1.3.).

One political theme to pay attention to is the politics of scale or multi-scalar politics, where actors in a discursive or legal struggle may jump scale to find venues or jurisdictions that are more likely to support their aim (Swyngedouw 1997). Legal disputes are very common in the energy space, and choice of state or federal legal or legislative efforts are often reflections of efforts to play to power asymmetries. When California passed a law called the low-carbon fuel standard (LCFS), they were sued

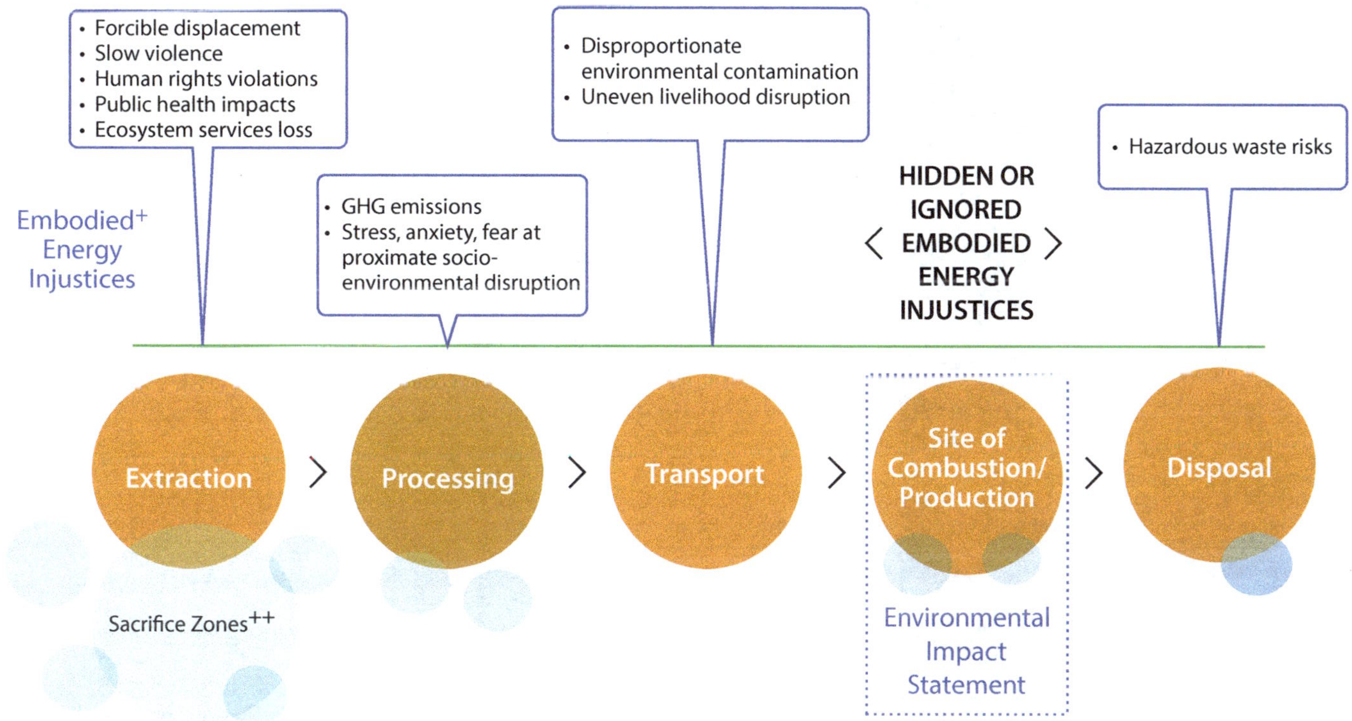

+ The injustices listed can occur anywhere along the supply-chain but typically are most prevalent around sites of extraction.
++ Sacrifice zones are areas poisoned or destroyed for the supposed greater good of economic progress.

Fig. 1.3 Embodied energy injustice (Healy et al. 2019)

Fig. 1.4 The Navajo generating station near Page, Arizona, brought about US$30 million in revenues to Navajo Nation annually

by corn farmers in the Midwest, arguing the state policy violated the federal Commerce Clause of the Constitution, which says states cannot regulate the practices of other states. The effort eventually failed, but had the effort worked, the power of the corn growers would have prevailed over the GHG laws of California.

There is a major rift in these discussions, owing to the technical nature of some of the questions in energy transitions. Science and Technology Studies (STS) reminds us that we need to treat these models and scenarios with humility, understanding that even our best models of nature and human interactions with technology cannot be easily predicted. Other STS scholars point to the power and performance of these technical pathway scenarios and models (Svetlova 2012). They become "black-boxes" that groups need to defend (Fig. 1.4.).

It is a matter of debate in some circles about to what degree energy transitions need to invoke ideas about justice, but it is this author's contention that they do. A just transition should focus on three pillars: energy access, providing support to those impacts by the fossil and conventional energy economy, and managing any impacts to communities and landscapes that might be caused by energy transitions (Newell & Mulvaney 2013). Deep decarbonization strategies might find this framework of three pillars of energy justice a useful check in thinking about our energy futures. First, pursuing energy strategies that ensure energy access for those who do not have it. Second, justice for those who work within and are affected by the fossil fuel economy, such as those living near power plants and industrial facilities, sometimes called fence-line communities. Third, to manage the potential impacts from pursuing decarbonization and climate justice, meaning any impacts that might arise from renewable energy or climate adaptation (Fig. 1.5.).

Fig. 1.5 Might Moss Landing Power Plant was once California's largest and soon will be home to its largest lithium-ion battery

References

Adger, W. N., Benjaminsen, T. A., Brown, K., & Svarstad, H. (2001). Advancing a political ecology of global environmental discourses. *Development and Change, 32*(4), 681–715.

Akin, M., & Urpelainen, J. (2018). *Renewables: The politics of a global energy transition*. Cambridge, MA: MIT Press.

Araújo, K. (2014). The emerging field of energy transitions: Progress, challenges, and opportunities. *Energy Research & Social Science, 1*, 112–121.

Ayres, R. W. (1975). Policing plutonium: The civil liberties fallout. *Harvard CR-CLL Review, 10*, 369–444.

Berman, D. M., & O'Connor, J. T. (1996). *Who owns the sun?* Burlington, VT: Chelsea Green Publishing Company.

Bloch, H., Rafiq, S., & Salim, R. (2015). Economic growth with coal, oil and renewable energy consumption in China: Prospects for fuel substitution. *Economic Modelling, 44*, 104–115.

BP. (2019). Statistical review of world energy. London, 69th Edition. bp.com/statisticalreview.

Brundtland, G. (1987). Our common future: Report of the 1987 world commission on environment and development. *United Nations, Oslo, 1* 59.

Bulkeley, H., McGuirk, P. M., & Dowling, R. (2016). Making a smart city for the smart grid? The urban material politics of actualising smart electricity networks. *Environment and Planning A: Economy and Space, 48*(9), 1709–1726.

Calvert, K. (2016). From 'energy geography' to "energy geographies": Perspectives on a fertile academic borderland. *Progress in Human Geography, 40*, 105–125.

Chappells, H., & Shove, E. (2005). Debating the future of comfort: Environmental sustainability, energy consumption and the indoor environment. *Building Research & Information, 33*(1), 32–40.

Clack, C. T., Qvist, S. A., Apt, J., Bazilian, M., Brandt, A. R., Caldeira, K., & Jaramillo, P. (2017). Evaluation of a proposal for reliable low-cost grid power with 100% wind, water, and solar. *Proceedings of the National Academy of Sciences, 114*(26), 6722–6727.

Delucchi, M. A., & Jacobson, M. Z. (2011). Providing all global energy with wind, water, and solar power, part II: Reliability, system and transmission costs, and policies. *Energy Policy, 39*(3), 1170–1190.

Geels, F. W. (2004). From sectoral systems of innovation to socio-technical systems: Insights about dynamics and change from sociology and institutional theory. *Research Policy, 33*(6–7), 897–920.

Geels, F. W., Sovacool, B. K., Schwanen, T., & Sorrell, S. (2017). Sociotechnical transitions for deep decarbonization. *Science, 357*(6357), 1242–1244.

Grady-Benson, J., & Sarathy, B. (2015). Fossil fuel divestment in US higher education: Student-led organizing for climate justice. *Local Environment, 21*(6), 661–681.

1

Grubler, A., Wilson, C., Bento, N., Boza-Kiss, B., Krey, V., McCollum, D. L., & Cullen, J. (2018). A low energy demand scenario for meeting the 1.5° C target and sustainable development goals without negative emission technologies. *Nature Energy, 3*(6), 515–527.

Haraway, D. (2015). Anthropocene, Capitalocene, Plantationocene, Chthulucene: Making kin. *Environmental Humanities, 6*, 159–165.

Hawken, P., Lovins, A. B., & Lovins, L. H. (2013). *Natural capitalism: The next industrial revolution*. London: Routledge.

Healy, N., Stephens, J., & Malin, S. (2019). Embodied energy injustices: Unveiling and politicizing the transboundary harms of fossil fuel extractivism and fossil fuel supply chains. *Energy Research & Social Science, 48*, 219–234.

Hess, D. (2012). *Good green jobs in a global economy: Making and keeping new industries in the United States*. Cambridge, MA: MIT Press.

Hondo, H., & Baba, K. (2010). Socio-psychological impacts of the introduction of energy technologies: change in environmental behavior of households with photovoltaic systems. Applied Energy, 87(1), 229–235.

International Energy Administration. (2016). *World energy outlook: A forward look to 2040*. Paris, France. Universe Books, New York.

International Energy Agency. (2010). *World energy outlook*. https://www.iea.org/reports/world-energy-outlook-2010.

Jacobson, M. Z., & Delucchi, M. A. (2009). A path to sustainable energy by 2030. *Scientific American, 301*(5), 58–65.

Jacobson, M. Z., & Delucchi, M. A. (2011). Providing all global energy with wind, water, and solar power, part I: Technologies, energy resources, quantities and areas of infrastructure, and materials. *Energy Policy, 39*(3), 1154–1169.

Jacobson, M. Z., Delucchi, M. A., Bauer, Z. A., Goodman, S. C., Chapman, W. E., Cameron, M. A., & Erwin, J. R. (2017). 100% clean and renewable wind, water, and sunlight all-sector energy roadmaps for 139 countries of the world. *Joule, 1*(1), 108–121.

Jacobson, M. Z., Delucchi, M. A., Cameron, M. A., & Frew, B. A. (2017). The United States can keep the grid stable at low cost with 100% clean, renewable energy in all sectors despite inaccurate claims. *Proceedings of the National Academy of Sciences, 114*(26), E5021–E5023.

Jenkins, K. (2018). Setting energy justice apart from the crowd: Lessons from environmental and climate justice. *Energy Research & Social Science, 39*, 117–121.

Jiusto, S. (2009). Energy transformations and geographic research (pp. 533–551). Blackwell Publishing, Oxford.

Juisto, S. (2009). Energy transformations and geographical research. In N. Castree, D. Demeritt, D. Liverman, & B. Rhoads (Eds.), *A companion to environmental geography* (pp. 533–551). Hoboken, NJ: Wiley-Blackwell.

Kallis, G. (2011). In defense of degrowth. *Ecological Economics, 70*(5), 873–880.

Langlois, K. (2018). Tiny bits of plastic permeate our world. *High Country News*, September 10, 2018. https://www.hcn.org/issues/50.20/pollution-tiny-bits-of-plastic-permeate-our-world.

Leach, G. (1992). The energy transition. *Energy Policy, 20*(2), 116–123.

Loftus, P. J., Cohen, A. M., Long, J. C. S., & Jenkins, J. D. (2014). A critical review of global decarbonization scenarios: What do they tell us about feasibility? *Wiley Interdisciplinary Reviews: Climate Change, 6*(1), 93–112.

Lombard, A., & Ferreira, S. L. (2015). The spatial distribution of renewable energy infrastructure in three particular provinces of South Africa. *Bulletin of Geography. Socio-Economic Series, 30*(30), 71–86.

Lovins, A. B. (1976). Energy strategy: The road not taken. *Foreign Affairs, 55*, 65–96.

Lucas, A. (2016). Stranded assets, externalities and carbon risk in the Australian coal industry: The case for contraction in a carbon-constrained world. *Energy Research & Social Science, 11*, 53–66.

McKibben, B. (2012). Global warming's terrifying new math. *Rolling Stone*, August 2, 2012. https://www.rollingstone.com/politics/politics-news/global-warmings-terrifying-new-math-188550/.

Meadows, Donella H., Dennis L. Meadows, Jorgen Randers, William W. Behrens III (1972). The limits to growth: A report to the club of Rome. Universe Books, New York.

Merkley, J. (2017). Merkley, Sanders, Markey, Booker introduce landmark legislation to transition United States to 100% clean and renewable energy. Press Release. Thursday, April 27, 2017.

Miller, C. A., & Richter, J. (2014). Social planning for energy transitions. *Current Sustainable/Renewable Energy Reports, 1* 77–84. https://doi.org/10.1007/s40518-014-0010-9.

Milne, M. J., & Gray, R. (2012). W(h)ither ecology? The triple bottom line, the global reporting initiative, and corporate sustainability reporting. *Journal if Business Ethics, 118*, 13–29.

Moore, J. (2015). *Capitalism in the web of life*. New York: Verso.

Newell, P., & Mulvaney, D. (2013). The political economy of the 'just transition'. *The Geographical Journal, 179*(2), 132–140.

Notter, D. A., Meyer, R., & Althaus, H. J. (2013). The Western lifestyle and its long way to sustainability. *Environmental Science & Technology, 47*(9), 4014–4021.

Platts. (2017). S&P global market intelligence world electric power database.

Pyne, S. (2015). The fire age: We can melt ice sheets and cook landscapes. When humans made fire, they made themselves and their planet too. https://aeon.co/essays/how-humans-made-fire-and-fire-made-us-human.

Ruddiman, W. F. (2018). Three flaws in defining a formal 'Anthropocene'. *Progress in Physical Geography: Earth and Environment, 42*(4), 451–461.

Ryghaug, M., Skjølsvold, T. M., & Heidenreich, S. (2018). Creating energy citizenship through material participation. *Social Studies of Science, 48*(2), 283–303.

Sadler, D., & Lloyd, S. (2009). Neo-liberalising corporate social responsibility: A political economy of corporate citizenship. *Geoforum, 40*(4), 613–622.

Schumacher, E. F. (1973). *Small is beautiful: Economics as if people mattered.* New York: Harper & Row.

Smil, V. (2004). *World history and energy. Berkshire Encyclopedia of World History*. New York: Elsevier.

Smil, V. (1984). On energy and land: Switching from fossil fuels to renewable energy will change our patterns of land use. *American Scientist, 72*(1), 15–21.

Smil, V. (2010). *Energy transitions: History, requirements, prospects*. ABC-CLIO: Santa Barbara.

Sovacool, B. (2013). *Energy and ethics: Justice and the global energy challenge*. Springer. London.

Sovacool, B., & Scarpaci, J. (2016). Energy justice and the contested petroleum politics of stranded assets: Policy insights from the Yasuní-ITT Initiative in Ecuador. *Energy Policy, 95*, 158–171.

Steffen, W., Rockström, J., Richardson, K., Lenton, T. M., Folke, C., Liverman, D., Donges, J. F. (2018). Trajectories of the Earth System in the Anthropocene. Proceedings of the National Academy of Sciences, 115(33) 8252–8259.

Steinbacher, K., & Röhrkasten, S. (2019). An outlook on Germany's international energy transition policy in the years to come: Solid foundations and new challenges. *Energy Research & Social Science, 49*, 204–208.

Svetlova, E. (2012). On the performative power of financial models. Economy and Society, 41(3), 418–434.

Swyngedouw, E. (1997). Neither global nor local: "Glocalization" and the politics of scale. *Space of Globalization: Reasserting the power of the local*, 115–136.

Swyngedouw, E., & Ernstson, H. (2018). Interrupting the anthropo-obScene: Immuno-biopolitics and depoliticizing ontologies in the Anthropocene. *Theory, Culture & Society, 35*(6), 3–30.

Truby, J. (2019). Decarbonizing bitcoin: Law and policy choices for reducing the energy consumption of Blockchain technologies and digital currencies. *Energy Research & Social Science, 44*, 399–410.

Turner, F. (2006). How digital technology found utopian ideology: Lessons from the first hackers' conference. *Critical Cyberculture Studies*, 257–269.

Wolf, J., & Moser, S. C. (2011). Individual understandings, perceptions, and engagement with climate change: Insights from in-depth studies across the world. *Wiley Interdisciplinary Reviews: Climate Change, 2*(4), 547–569.

White, D. F. (2019). Ecological democracy, just transitions and a political ecology of design. *Environmental Values, 28*(1), 31–53.

Whyte, K. (2017). The Dakota access pipeline, environmental injustice, and US colonialism.

Winner, L. (1986). Do artifacts have politics? The whale and the reactor: a search for limits in an age of high technology. Univeristy of Chicago Press, Chicago.

Winner, L. (1989). *The whale and the reactor: A search for limits in an age of high technology*. Chicago: University of Chicago Press.

Yates, J. J. (2012). Abundance on trial: The cultural significance of sustainability. *The Hedgehog Review, 14*(2), 8–25.

Zweibel, K., Mason, J., & Fthenakis, V. (2008). A solar grand plan. *Scientific American, 298*(1), 64–73.

Fundamentals of Energy Science

Contents

D. Mulvaney, *Sustainable Energy Transitions*,
https://doi.org/10.1007/978-3-030-48912-0_2

2

Learning Goals

After this chapter, readers will be able to

- describe the physical principles related to energy, heat, power, and work;
- complete basic calculations and unit conversions in energy, heat, power, and work;
- describe the scientific properties of conventional and renewable energy sources; and
- appreciate the scale and magnitude of energy transformations.

Overview

This chapter reviews keywords, concepts, and calculations used in the energy sciences. The first entry point into conversations in the energy sector is an understanding of the underlying physics, chemistry, and other key concepts, definition, principles, and metrics used in energy science. The chapter reviews how energy flows from the sun through nature and time, emphasizing how energy is transferred through Earth's climate, ecosystems, and what geologist Eduard Suess called the anthrosphere—the areas of the Earth transformed by humans (Ellis 2008). Key ideas covered in this chapter include energy types and sources, energy conversions and transformations, laws of thermodynamics, and other key scientific principles and descriptions of infrastructure needed to understand energy systems. Throughout this chapter, there are example problems that review energy-to-power relations, unit conversions, and stoichiometry, in addition to a handful of empirical problem sets and assignment. The chapter ends with a discussion about how concepts like entropy can provide a scientific basis for understanding how natural resource use has implications for intergenerational equality—the idea that resources must be managed for future generations (Costanza et al. 2014).

2.1 Power and Energy

Energy is fundamental stuff. It is what allows life to occur, through metabolism and eventually decay. It shapes where non-living things are and where they are going. As the property of a star, the amount of energy determines whether humans could eventually inhabit those places. Energy also enables humans to develop the societies that exist today, and the linkages between peoples, economies, and nature in different parts of the world. Vaclav Smil describes energy use by human civilization as its metabolism—borrowing a metaphor for how living organisms process energy to fuel their needs and wants.

What is energy? From your biology, physics, or chemistry classes in school you may recall that energy is the ability to do work or the ability to transform a system. Energy is encountered in different forms, but most often its presence is suggested by motion, activity, light, heat, or change.

As a fundamental law of the universe, energy is always conserved. The energy that was present at the start of the universe is all still in the universe. This means energy cannot be created, only converted and transformed. Energy is a discrete quantity. This makes it different than power, which is a flow rate quantity of energy. Energy is the amount of power over time measured in joules (J). Power is the amount of energy released at a specific time, like a joule per second (1 watt).

Units to measure energy besides the joule include some that you may encounter in everyday life: British thermal units (Btu), kilo watt hours (kWh), quads (a quadrillion Btu), kilo calories, or therms (th). One Btu is about the amount of energy contained in a single match stick, while 100 quadrillion Btu of energy is about the amount used annually in the US. A joule is one match strike divided into 1,055 parts. A unit to measure explosive energy is "TNT equivalent" and the unit is equal to 4.184×10^9 J. Nuclear explosions are measured in megatons, officially designated to be equal to 1 million tons of TNT equivalent. These units are important to help us understand the magnitude and size of our energy systems and their components and subsystems. As the late energy analyst David MacKay (2015) argues, numbers and their importance cannot be understated. "Numbers are chosen to impress, to score points in arguments, rather than to inform" (MacKay 2015, 3). Our goal is to explore how numbers can be put into meaningful context, appreciate their magnitude, and become familiar with their units.

Many energy conversations benchmark power use by energy devices to the 100-watt incandescent lightbulb. But that is becoming increasingly anachronistic as light-emitting diode (LED) bulbs replace the conventional incandescents, and efficiency in incandescent bulbs has improved. One 100-watt incandescent can be replaced with one 8- to 12-watt LED and provide the same amount of light. The brightness of visible light for humans is measured in lumens and candle-feet, so the power use is less important to the consumer than the brightness or lumens. Lighting has been a major focus of energy efficiency improvements for many decades, with large energy savings in the 1970s as fluorescent lights were adopted.

Another device in use in a typical household in a developed country that uses about 100 watts is a flat-panel television screen. Older devices may be less efficient, or the screen may not be in the low-power mode by default, but the average of new televisions is on the order of 100 watts. An order of magnitude lower—10 watts—is closer to what a cell phone charger might use to "charge up" your phone. A hairdryer or microwave is usually measured in the 1,000-watt level, with some devices at higher levels. Other important items that draw a lot of power include an electric car, using over 7,000 watts. An electric hot water heater can use nearly 5,000 watts while heating water. A heat pump water heater will use about three times fewer watts (but they come in all sizes).

So far, the upshot is that energy is a discrete quantity that provides heat, motion, and light. Energy occurs in many forms, including gravitational potential energy, kinetic energy, electromagnetic (EM) radiation, heat, and beta and gamma rays. The energy contained in gasoline is stored in the chemical bonds that hold gasoline molecules together. This is why when fuels like gasoline or natural gas are combusted, they liberate energy in the form of heat, flame, light, and sound. We make gasoline, diesel fuel, and electricity to move ourselves and our things around. The combustion of gasoline is converted into wheel motion by transferring the work done by heat into wheel motion. Electric motors turn power into motion in an electric car. We move electricity along filaments to produce light in lightbulbs to light our homes. We can burn natural gas, propane, or biogas on-site, deliver electricity, or passively utilize the sun to heat a home. Each of these processes utilize one or more of the forces when systems are transformed: gravitational, electromagnetic, and two nuclear forces.

What are forces? Forces are the ultimate drivers of interaction between two or more objects. Energy makes it possible to transform systems based on rules of four forces. The gravitational force governs the large-scale structure of the universe. Transferring gravitational potential energy to kinetic power makes it possible to generate electricity with hydropower. The electromagnetic force is responsible for the structure of matter at the molecular level. Chemical reactions between molecules during combustion liberate heat from chemical bonds, and this heat makes engines and steam turbines work. The strong and weak nuclear forces govern the structure of matter at the subatomic level—they are the forces that bond together protons and neutrons. The release of this energy from nuclear fission is the driving force behind nuclear power plants that generate electricity.

Definition

Power is the rate flow of energy. It is an amount of instantaneous energy flow. This means that power has the unit dimension of time. A watt is also described as 1 joule per second.

A related measure of power is horsepower (hp)—1 hp = 745.7 watts. James Watt used a coal mine pony as the benchmark to define how much work could be done by a small horse in a coal mine. Work is the amount of force per unit distance and time. The amount of work capable by a small horse working a mine was about 22,000 foot-pounds of work in a minute. There are several ways to comprise this metric but imagine 220 pounds of coal, pulled 100 feet in 1 minute. That defines the modern metric horsepower.

Hydrocarbons such as methane (CH_4) and propane (C_3H_8) when reacted with oxygen (O_2) produce carbon dioxide (CO_2) and water (H_2O). Methane can be produced by fossil or biogenic sources. Methane in the form of dry natural gas is used to heat homes and power fuel cells. It is generated with a mix of fossil natural gas and biogenic sources in landfills due to the decomposition of organic materials. Similarly, wastewater treatment facilities, dairies, and other animal food systems can be sources of methane generation. Because methane is a potent greenhouse gas (GHG), each molecule releases about 25 times more GHG pollution, so many mitigation strategies aim to convert CH_4 to CO_2.

Definition

Stoichiometrically, methane combustion produces carbon dioxide, water, and heat.

$$CH_4 + 2O_2 \rightarrow CO_2 + 2H_2O + \text{heat}$$

Below is the chemical reaction for combustion of ethanol, a biofuel produced mainly from corn and sugarcane, and often blended with gasoline. Other feedstocks used to make ethanol include wheat, sugar beet, algae, miscanthus, and switchgrass—essentially anything with biomass that can be fermented. The easiest have fermentable sugars (sugarcane, sugar beet) or starches that can be easily converted to fermentable sugars.

Definition

Combustion of ethanol is an exothermic reaction and yields heat, water, and CO_2.

$$C_2H_5OH + 3O_2 \rightarrow 2CO_2 + 3H_2O + \text{heat}$$

Note that the molecular weight of carbon dioxide (carbon = 12 + 2 oxygen molecules at 16 each, equals 44) is significantly higher than methane (carbon = 12, 4 hydrogens at 4 each brings the total to 16). This means that the GHG emissions are more than double the initial mass of fuel.

Some disciplines and research clusters use carbon dioxide equivalent (CO_2e), while others use carbon (C) as a metric for emissions. Use the 44/16 ratio to make the necessary conversions between C and CO_2e.

The following table describes the size and scale of output for various energy sources used to generate electricity.

Definition

The scale of energy output for different energy sources for electricity

Energy facility	Country	Annual output (GWhs)
Three Gorges Dam	China	87,000
Palo Verde Nuclear Power Plant	US (Arizona)	32,000
Hoover Dam	US (Nevada)	4000
Topaz Solar Farm	US (California)	1300
Hatchōbaru Geothermal Power Plant	Japan	785
Navajo Coal-Generating Station	US (Arizona)	12.5
Muppandal Wind Farm	India	10

Example 2.1 Understanding the Relationship Between Units of Power and Energy

Consistent use of units is a convenient way to ensure your dimensional analysis checks out. If the units are canceled correctly, it is a good way to confirm your calculations are correct.

1. **A 60-watt incandescent lightbulb in an attic is left on all day and night for every day for a year. (a) How much energy does this consume in kWh? (b) If the incandescent lightbulb was replaced with an 8-watt LED of equivalent lumens and put on a timer to be on 8 hours per day, how much energy would it consume in kWh? (c) What is the energy savings from this change?**
   (a) Convert watts to kilowatts and multiply by the number of hours in a year, since the device is drawing power all year.
   $$60\,\mathrm{W} \times \frac{1\,\mathrm{kW}}{1000\,\mathrm{W}} \times \frac{24\,\mathrm{h}}{\mathrm{d}} \times \frac{365\,\mathrm{d}}{\mathrm{y}} = 525.6\,\mathrm{kWh}$$
   (b) Now, estimate the energy used in the same setup except with the lower power LED and now only 8 hours per day.
   $$8\,\mathrm{W} \times \frac{1\,\mathrm{kW}}{1000\,\mathrm{W}} \times \frac{8\,\mathrm{h}}{\mathrm{d}} \times \frac{365\,\mathrm{d}}{\mathrm{y}} = 23.4\,\mathrm{kWh}$$
   (c) This swap of lightbulbs and adding a timer can yield a pretty big savings!
   525.6 kWh – 23.4 kWh = 502.2 kWh saved from an energy efficiency upgrade.
2. **AAA batteries contain 1.41 Wh of energy. If a brand-new battery can power a headlamp for 32 hours, how much power (in watts) does the lamp use during operation?**
   $$\frac{1.41\,\mathrm{Wh}}{32\,h} = 0.044\,\mathrm{W}$$

Keep an eye out for the many different units we use to measure power and energy as you read articles and news stories (Fig. 2.1).

Fig. 2.1 Watt's 100 kW steam engine versus modern steam turbine, which can be as large as 1.5 GW, or a million times more powerful

2.2 Electromagnetic Induction and Electricity

Electricity is a high-quality form of energy that is highly versatile and capable of tasks that other energy forms are not. It can provide many different services, is efficient in transferring power to motion, and is relatively easy to transport along wires. It can also store charges in transistors to make computers work. Electricity is the highest quality of energy able to produce many things from heat to operating complex computers. But how is electricity made?

To understand this requires we revisit Ampere's law and the laboratory work of Michael Faraday. Electricity is created from any process that forces electric charges apart. In a battery with electrodes at the poles that are by convention labeled positive—the anode—and the other negative—the cathode—the chemical composition of the battery is what forces charges apart. As the battery is depleted, the charge separation processes across this electric field lose their effectiveness, and the battery loses power. If the battery is a rechargeable one, external electricity can be added to restore the charge. Generators, batteries, fuel cells, and photovoltaic modules all utilize electric fields to make electricity.

Distance between two charges affects electricity in an analogous way to how gravity affects two separate body masses, such as the Earth and the Moon. The force of attraction in gravity between two bodies is represented by Sir Isaac Newton as the universal law of gravitation

$$F = \frac{GmM}{r^2}$$

Here, G is the force of gravity, while m and M are the mass of each of two respective objects. In the denominator, r is the distance between the two objects. As the masses are pulled apart, they exert less force on each other.

Whereas the force (F) of charge separation is represented as

$$F = \frac{kqQ}{r^2}$$

Here, k represents the electric constant, and the Q and q are the charges of the two objects. Once again, r is the distance between the objects. The strength of electric fields is characterized by the electric force per unit of charge. The electric force between objects is analogous to how the force of gravity relates across distance and mass. In each, force is inversely proportional to distance. It gets weaker as the distance grows.

The key relationship to remember with electricity is that electric fields cause magnetic fields, and vice versa: magnetic fields produce electric currents. These magnetic and electric fields are present in fluxes that operate perpendicular to each other and in a direction according to the right-hand rule. In other words, the rotation of the magnetic field's flow of force moves counterclockwise to the direction of electric current flow.

Another critical term used in energy studies is voltage, a measure of electrical potential. The units for voltage are volts, or broken down further as joules per coulomb—energy per unit charge. The electric current equals the flow of electric charge, typically carried by electrons and measured in amperes or amps.

Electromagnetic induction is a critical principle that underlies much of modern electricity generation and transmission infrastructure. Most electricity power plants operate by induction using steam or gas turbines, so do transmission and distribution equipment such as transformers. This basically means that the rotation of a magnetic field can induce an electric field. The physicist Michael Faraday and others experimented with induction using batteries as the supply of energy and galvanometers to measure electric current. The experiments used two coils of wire, one connected by wire conductors to a battery. The second coil had no current but was connected to the galvanometer. When the battery-connected coil was dropped inside the slightly larger non-electrified coil, the galvanometer registered the presence of electric current. The electricity moving through the battery-powered coil generated a magnetic field and the magnetic field stimulated electric current in the wire coil. Energy efficiency experts contend that induction stoves are superior technologies at delivering heat to cooking pans to natural gas or propane stoves, and far superior to electric heating stoves.

Definition

Electric current is the flow of electricity through a conductor.

Electric current produces a magnetic field perpendicular to the direction the current is moving in loops that travel according to the right-hand rule. The right-hand rule says that electric current flows in the direction that your thumb points when a magnetic field moves in the direction your other fingers point when you make a fist. In the experiment, when the electrified first coil is placed inside of the non-electrified second coil, the magnetic fields coming from the first coil produce electric fields in the second coil. The second coil is able to produce electricity because the magnetic fields induce the current. This current is measured in the galvanometer. This proved experimentally that electric fields produce magnetic fields, and magnetic fields produce electric fields, the law known as Ampere's law (◘ Fig. 2.2).

The rotation of a conductor coil across a magnetic field generates electrical current. This is the basic idea behind most of the devices that generate electricity with moving parts—natural gas turbines, steam turbines at nuclear power plants,

Fig. 2.2 Electromagnetic induction and the demonstration of Ampere's law by Michael Faraday

hydropower plants, concentrated solar power technologies, wind turbines, generators, automobile alternators, and geothermal power plants, just to name a few. The simplest version of this is called a dynamo, a device that consists of a coil of copper wire attached to a rotating shaft that contains a permanent magnet. The magnet provides the magnetic field, so when the shaft is rotated, an electric field is generated in the copper wire coil, and that electric current is drawn into a circuit—in the power plants mentioned earlier, this means the electricity grid in most cases. This rotation produces alternating current from positive to negative poles, as opposed to direct current, and most electricity networks do it at 60 cycles per second, and do it in three phases rotated at 120 degrees.

2.3 Laws of Thermodynamics

The first law of thermodynamics—the conservation of energy law—is that the total content of the universe is constant. The second law of thermodynamics—the entropy law—is that the total entropy is increasing. Entropy is a measure of the amount of energy no longer capable of conversion into work. The idea behind entropy is that useful energy is dissipated with each energy conversion—with each transformation, we take high-quality, low-entropy resources and convert them into high-entropy energy (low quality, because it is no longer capable of work). This quality of energy will continue to degrade or increase entropy into the future.

Definition

Entropy—the measure of useful energy in a closed system.

The idea of entropy is also useful when thinking about the metabolism of natural resources caused by human interactions with the biophysical world. Entropy captures the irreversible aspects of energy transformation. Spontaneity emerges out of an upward shift in entropy, from order of low entropy to the chaos of high entropy. Low entropy systems are capable of causing a lot of change in a system. A system with high entropy is highly unordered (or chaotic) and contains a lot of energy incapable of work. High entropy energy has little capacity for movement. Entropy is a measure of the free energy compared to the bound energy. Recollect from ► Chapter 1, the highest state of the total entropy in the universe is cosmological heat death.

There is an astrophysical truth to the contention that as the universe ages, its energy is moving toward high entropy, making less energy available to cause movement. That temporal movement involves a transfer of heat or energy. As heat transfer goes toward equilibrium—say a hot pan dissipating heat in a cool room—work done in the process is available to be captured. But once equilibrium is attained, everything is at the same temperature and more energy from outside the interacting elements needs to be imported to restore the elements to their original form. The story of entropy for sustainability is that if we design technologies and social relations to utilize energy as efficiently as possible with the chaotic, unorganized, dissipation of useful energy, there will be more for us in the future or it our stuff will take us less space. Though that also makes assumptions that will be explored in the next chapter.

A system containing a glass of water and an ice cube is an example of a system of high order and low entropy. Theoretically, it is possible to capture some of the energy across the temperature differential as the system trends toward equilibrium. At equilibrium, the water will have no ice. Energy is transferred from the water into the ice to melt it. The energy from the ice is transferred into the water to cool it.

These biophysical realities set limits to our capacity to use the Earth as a natural resource. Nicholas Georgescu-Roegen introduced the idea into the study of economics in 1971 (Georgescu-Roegen 1971). He described low entropy energy as occurring in two forms. The first are stocks of accumulated solar energy or other natural resources. These are in limited quantity. The second are flows, which are constantly replenishing energy sources. This distinction is important because it suggests that stocks might eventually run out. If we know the stock of a given energy resource, for example, and we know the rate of depletion, we can estimate the number of years until no more low entropy stocks are left. This could be a long time with some resource, but less in other instances.

This is an important observation—one that is foundational to ecological economics—because it reveals that the economy neither produces nor consumes matter or energy; it only absorbs matter-energy and converts it to waste continuously. This suggests that economy from a resource perspective is linear, not circular. In nature, there are water cycles and nutrient cycles, and wastes become resources. The idea of a circular economy is to move beyond the "produce, use, dispose" mentality and design the economy toward that of wastes as resources for others (Stahel 2016).

2.4 Exercise

2.4.1 Basic Unit Conversions

Questions

1. Convert 210 kWh into joules (J).
2. How many kWh are there in 101,000 Btu?
3. Convert 150 kilocalories into giga-joules (GJ).
4. Ten gallons of gasoline contains how much energy (MJ)?
5. The US used approximately 102 quads of energy in 2018. Convert this to (a) TWh (terawatt-hours) and (b) EJ (exajoules).
6. A household uses about 6,721 kWh per year. What is the annual energy consumption of an average household expressed in (a) tons of coal equivalent (tce)? (b) ton of oil equivalent (toe)?

Answers to these questions are at the end of the chapter.

Exergy is another helpful thermodynamic concept for thinking about the foundational physics of energy transitions from the second law. It refers to the amount of useful energy available to do work relative to the system. This measure of energy quality can be a useful way to help think about the efficient use of energy since it is a measuring of how much useful work can be obtained through an energy transformation with the minimal loss of entropy. A device that is inefficient is said to have a low exergy output, whereas an energy-efficient device can have a high exergy output. While energy cannot be created or destroyed, exergy *can be* destroyed because it is a measure of energy's potential and degradation. Irreversible processes—such as digging up coal and burning it—also destroy exergy. When a subsystem comes into equilibrium with its surroundings—a process that is irreversible unless external inputs are added—there is zero exergy. When a reservoir for a hydropower facility that depends on it for power generation falls below the level for its hydropower generation, it no longer has exergy until the reservoir is filled with water from an external system (say, rainfall, or release of water from an upstream dam) again.

2.5 Photon Science

For 4.4 billion years, our sun has generated photons as it balances the pressure of its weight from gravity against the outward push of energy release from the fusion of hydrogen. The tremendous weight of the sun's gravity momentarily produces a hydrogen isotope with an extra proton—two overall. When the atom relaxes back to the more common hydrogen with only one proton, it releases the energy in the form of light. The loss of the temporary proton accelerates a charge, which is where electromagnetic radiation from our sun originates.

2

The power from the sun is 3×10^{26} watts or 1.360 watts per square meter (NASA 2019). That is enough to power 300 trillion 100-watt laptop computers at one time. About 10^{17} watts is used by humans on Earth. Energy in the form of electromagnetic radiation interacts with matter in peculiar ways. The light interaction with matter can cause some light to be reflected, absorbed, or transmitted through the material. Also, the light can be temperature dependent, as black body radiation demonstrates when iron is placed in a fire.

Definition

The energy of a photon is represented as (E). Plank's constant (f) and the speed of light (c) are variables in the determination of energy, which ranges depending on the wavelength (λ) of light.

$$E = hf = \frac{hc}{\lambda}$$

As different photons have different energies, it is not clear whether any given electron has enough energy to excite the electron without knowing the bandwidth. Energy and bandwidth are inversely proportional. This is where the concept of a bandgap is critical to understanding the materials needed for photovoltaics. The band gap is the difference between available energy states for an electron. Some materials have better bandgaps corresponding to the sun's black-body radiation. If the photon has enough energy to raise the electron's energy above the bandgap, it will be able to move into the electric circuit. If there is not enough energy, the electron will relax (◘ Fig. 2.3). The Fermi energy is the hypothetical middle energy level, where is it 50% odds of finding the electron.

The science of color is quite spectacular given the diversity of species that "see" and the various different ways the light spectrum and matter interact (◘ Fig. 2.4).

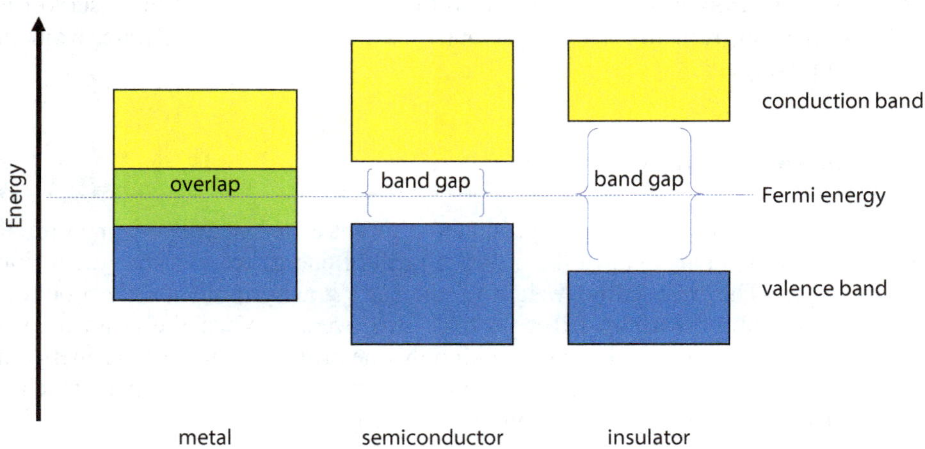

◘ **Fig. 2.3** Bandgaps of semiconductor materials used in photovoltaics

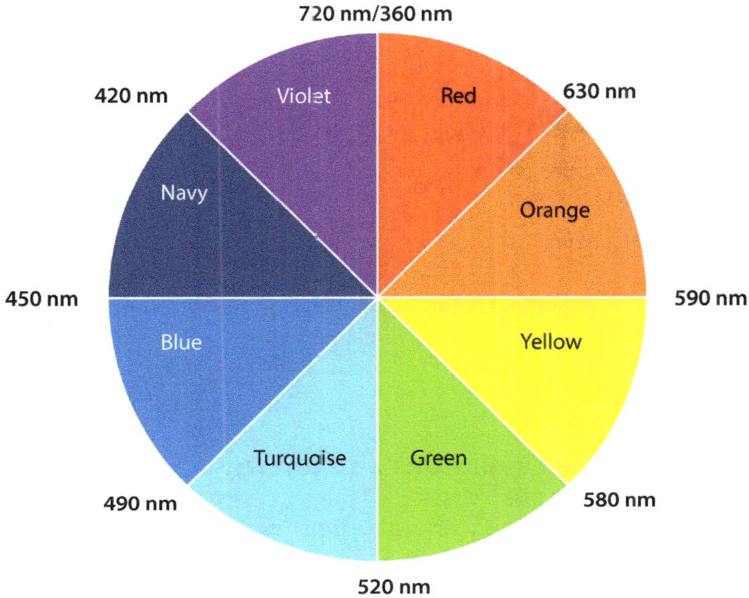

Fig. 2.4 Color wheel of light absorbed and color

Example 2.2 Energy in Photons and Interactions with Matter

Photons corresponding to the color blue (450 nm) have how much energy?

$$E = \frac{hc}{\lambda}$$

$$= \frac{6.6\times10^{-34}\,\frac{\text{m}^2}{\text{s}}\,\text{kg}\times2.9\times10^{8}\,\text{m/s}}{4.7\times10^{-7}\,\text{m}}$$

$$= 4.4\times10^{-19}\,\text{J}\,\text{or}\,2.75\,\text{eV}$$

Electromagnetic (EM) radiation of frequency (*f*) 3.1 × 10^{14} Hz (1/s) hits CdTe. The bandgap of CdTe is 1.44 eV. Does the EM have enough energy to excite an electron to the conductive band?

$$E = hf = \left(6.6\times10^{-34}\,\frac{\text{m}^2}{\text{s}}\,\text{kg}\right)\times\left(3.1\times10^{14}\,\text{s}^{-1}\right) = 2.04\times10^{-19}\,\frac{\text{m}^2}{\text{s}^2}\,\text{kg} = 2.04\times10^{-19}\,\text{J}$$

A semiconductor material is yellow-orange in color, which corresponds to a material absorbing all wavelengths below ~510 nm. What is the bandgap of this material? Using a figure in this chapter, can you speculate what the material actually is?

2

$$E = \frac{hc}{\lambda} = 2.43\text{eV}$$

$$\lambda = \frac{hc}{E} = \frac{6.626 \times 10^{-34} \frac{\text{m}^2}{\text{s}} \text{kg} \times 2.998 \text{x} 10^8 \text{m/s}}{2.43\text{eV} \times 1.6 \frac{\text{eV}}{\text{J}} \times 10^{-34}} = 5.1 \times 10^{-7} \text{ meter}(\text{m})$$

$$= 510 \text{ nanometers}(\text{nm})$$

At 510 nm, all wavelengths <510 nm (~green, blue, violet) are absorbed, so the compound is yellow-orange as shown in the color wheel.

2.6 Greenhouse Gas Emissions and Energy

It will be obvious to most readers that every energy source will have its own environmental profile per unit energy delivered. GHG emissions are one major difference among energy choices for energy systems. Combusting a gallon of gasoline emits about 20 pounds of CO_2, while combusting 100 gallons of gasoline emits one metric ton of CO_2. The ratio of carbon to hydrogen in fuels affects the amount of GHG emissions associated with each unit of energy from fossil fuels. For methane, the ratio of carbon to hydrogen is highest at 1:4. This is why natural gas emits fewer GHGs per unit energy than coal where the ratio of carbon to hydrogen is about 2:1, depending on the grade of coal. Most petroleum has a ratio of about 1:2. The energy contained in these fuels is bound up in chemical bonds that hold the molecule together. The carbon emission from coal per unit energy is highest, whereas natural gas has the lowest emissions per unit energy.

Speaking of GHG emissions, it is important to note that some analysts distinguish between direct and indirect emissions, or combine them and call them life-cycle emissions. Direct emissions are those that occur on-site, such as the carbon in the fuel that is liberated when fossil fuels are combusted. Indirect emissions include those up- or downstream of production. Taken together, direct and indirect emissions are often referred to as life-cycle emissions. Considering also the amount of energy and carbon needed to produce energy, the life-cycle emissions for natural gas, petroleum, and coal are 0.91 kg CO_2e/kWh, 0.77 kg CO_2e/kWh, and 0.51 kg CO_2e/kWh, respectively (◘ Fig. 2.5).

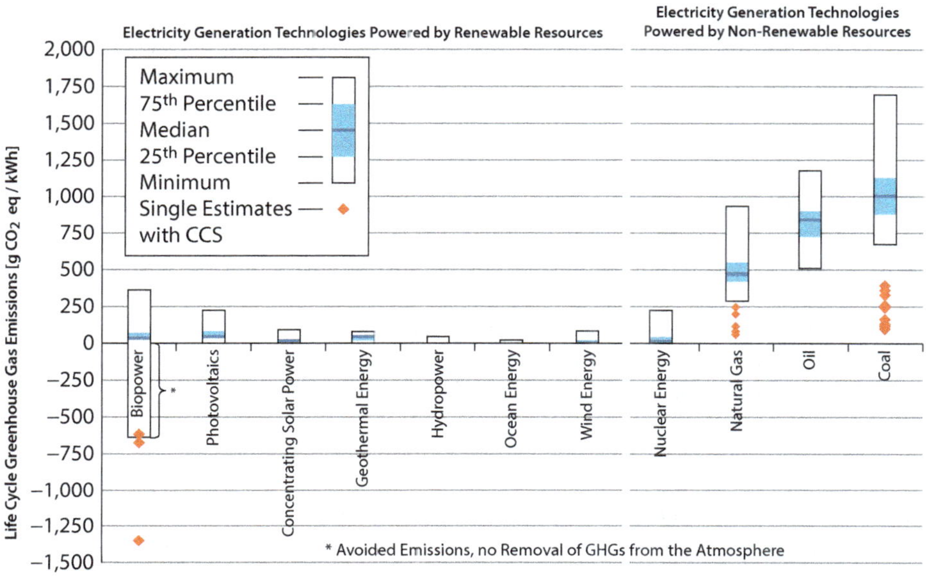

	Biopower	Photovoltaics	Concentrating Solar Power	Geothermal Energy	Hydropower	Ocean Energy	Wind Energy	Nuclear Energy	Natural Gas	Oil	Coal
Count of Estimates	222(+4)	124	36	8	28	10	126	125	83(+7)	24	169(+12)
Count of References	52(+0)	26	10	6	11	5	49	32	36(+4)	10	50(+10)

Fig. 2.5 GHG emissions factors for different sources of energy

2.7 Exercise

2.7.1 Dimensional Analysis

Questions

1. Replacing nuclear power with human power

   To appreciate the amount of energy that a large power plant can produce, let us compare its output to the number of humans that would be needed to replace it. Human beings are capable of doing approximately 100 watts of work per person. Diablo Canyon Nuclear Power Plant is capable of 2.35 GW of power.

   (a) How many human beings (turning a crankshaft) would it take to generate 2.35 GW of power? Assume that 50% of this power is converted to useful energy (meaning the crankshaft is 50% efficient at converting human power from kinetic energy into electricity).

   (b) If these workers were paid US$8 per hour, how much would the electricity cost in $/kWh?

   (c) If an average home consumes 500 kWh per month, how much would one month's electricity cost (in $)?

2

2. Replacing solar power with human power

   An athlete is capable of doing approximately 125 watts of work. A solar array made of 100 photovoltaic modules is able to produce 30 kW or 30,000 watts of power.

   (a) How many human beings (turning a crankshaft) would it take to generate 30 kW of power? Assume that 80% of this power is converted to useful energy (meaning the crankshaft is 80% efficient at converting human power into electricity).

   (b) If the solar array had a capacity factor (the percentage of time the solar array is generating power) of 0.25 or 25%, what would the daily electricity output be in kWh? If these workers were paid US$8 per hour for 6 hours per day (corresponding with the capacity factor), how much would the electricity cost in $/kWh?

Answers to these questions are at the end of the chapter (Fig. 2.6).

With some of these energy fundamentals down, we can move on to another set of important areas to understand energy transition: the social sciences.

2.8 Exercise Answers

2.8.1 Basic Unit Conversions

Answers

1. $210\,\text{kWh} \times \dfrac{3.6\,\text{MJ}}{1\,\text{kWh}} \times \dfrac{1{,}000{,}000\,\text{J}}{1\,\text{MJ}} = 756{,}000{,}000\,\text{J}$

2. $101{,}000\,\text{Btu} \times \dfrac{1\,\text{kWh}}{3{,}412\,\text{Btu}} = 29.6\,\text{kWh}$

3. $150\,\text{kcal} \times \dfrac{4.187\,\text{KJ}}{1\,\text{kcal}} \times \dfrac{1\,\text{GJ}}{1{,}000{,}000\,\text{KJ}} = 0.00063\,\text{GJ}$

4. $10\,\text{gallons gasoline} \times \dfrac{121.3\,\text{MJ}}{1\,\text{gallon gasoline}} = 1{,}213\,\text{MJ}$

5. $102\,\text{quads} \times \dfrac{293\,\text{TWh}}{1\,\text{quad}} = 29{,}886\,\text{TWh},$

   OR $102\,\text{quads} \times \dfrac{1\,\text{EJ}}{0.95\,\text{quad}} = 107.4\,\text{EJ}$

6. (a) $6{,}721\,\text{kWh} \times \dfrac{1\,\text{tce}}{8{,}141\,kWh} = 0.83\,\text{tce}$

   (b) $6{,}721\,\text{kWh} \times \dfrac{1\,\text{tce}}{11{,}630\,kWh} = 0.58\,\text{toe}$

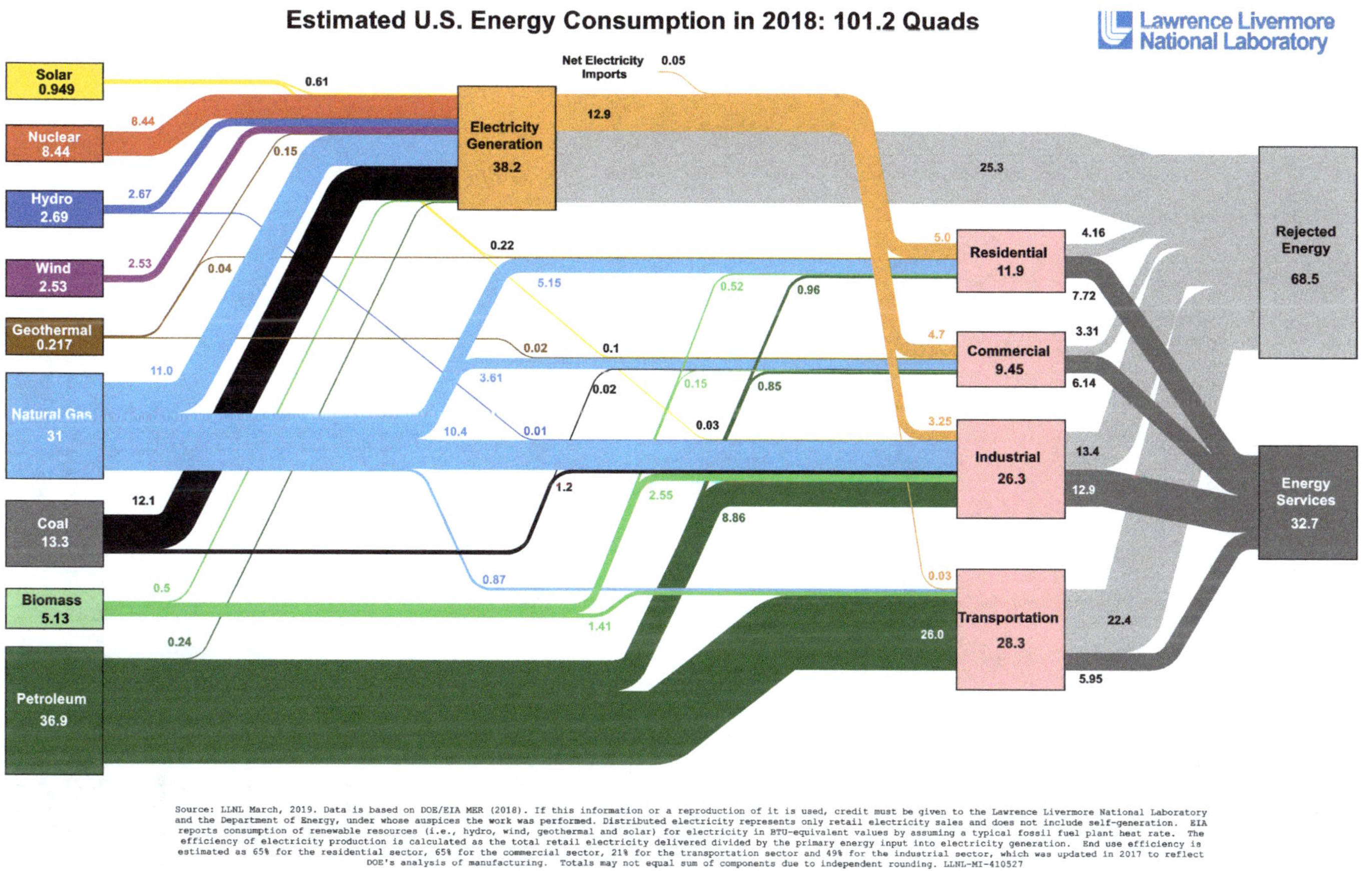

Fig. 2.6 US energy consumption in 2018 described by a Sankey Diagram

2

2.8.2 Dimensional Analysis

Answers

1. $$\frac{1\ \text{person}}{100\ \text{W}} \times \frac{1\text{W in}}{0.5\ \text{Wout}} \times \frac{2{,}350{,}000{,}000\ \text{W}}{\text{Nuclear power plant}} = \frac{47{,}000{,}000\ \text{people}}{\text{Nuclear power plant}}$$

$$2{,}350{,}000\ \text{kW} \times \frac{24\ \text{h}}{\text{d}} \times \frac{30\ \text{d}}{\text{month}} = \frac{1{,}692{,}000{,}000\ \text{kWh}}{\text{month}}$$

$$\frac{47{,}000{,}000\ \text{person}}{\text{shift}} \times \frac{3\ \text{shift}}{\text{d}} \times \frac{\$8}{\text{person}} \times \frac{30\ \text{d}}{\text{month}} = \frac{\$33{,}840{,}000{,}000}{\text{month}}$$

$$\frac{\frac{\$33{,}840{,}000{,}000}{\text{month}}}{\frac{1{,}692{,}000{,}000\ \text{kWh}}{\text{month}}} = \frac{\$20}{\text{kWh}}$$

US$20 per kWh is 100 times more than we pay for some of the more expensive electricity.

2. People needed to replace the power output of the solar array.

$$\frac{1\ \text{people}}{125\ \text{W}} \times \frac{1\text{W in}}{0.75\ \text{Wout}} \times \frac{25{,}000\ \text{W}}{\text{solar array}} = 266 \frac{\text{people}}{\text{solar array}}$$

Energy output of the solar array.

$$25\ \text{kW} \times \frac{24\ \text{h}}{\text{d}} \times 0.25 = 150\ \text{kWh/day}$$

Cost of energy from humans turning a crankshaft.

$$\frac{266\ \text{people}}{} \times \frac{6\text{h}}{\text{d}} \times \frac{\$8}{\text{h} * \text{person}} = \frac{\$12{,}768}{\text{day}}$$

$$\frac{\frac{\$12{,}768}{\text{day}}}{\frac{150\ \text{kWh}}{\text{day}}} = \frac{\$85}{\text{kWh}}$$

Compare US$85 per to the typical US$0.15 to US$0.45 price for electricity in the US.

References

Costanza, R., Cumberland, J. H., Daly, H., Goodland, R., Norgaard, R. B., Kubiszewski, I., & Franco, C. (2014). *An introduction to ecological economics*. Boca Raton, FL: CRC Press.

Ellis, E. (2008). The biosphere. The encyclopedia of the Earth. http://ecotope.org/people/ellis/papers/ellis_eoe_biosphere_2008.pdf.

Georgescu-Roegen, Nicholas (1971). The Entropy Law and the Economic Process Cambridge, Massachusetts: Harvard University Press. ISBN 978-0674257801.
MacKay, D. (2015). Sustainable energy without the hot air. https://www.withouthotair.com/.
Stahel, W. (2016). Circular economy. *Nature, 531*, 435–438.
U.S. National Aeronautics and Space Administration (NASA). (2019). Incoming sunlight. https://earthobservatory.nasa.gov/features/EnergyBalance/page2.php. Accessed 15 May 2019.

Energy and the Social Sciences

Contents

D. Mulvaney, *Sustainable Energy Transitions*,
https://doi.org/10.1007/978-3-030-48912-0_3

3

Learning Goals

At the end of this chapter, readers should have:

- a full appreciation of the contributions made by many different social science disciplines and leading scholars asking questions related to energy transitions;
- selected theories of social change critical to energy transitions;
- understand the structure and behavior of socio-environmental systems;
- consider the importance of scale and context in addressing socio-environmental problems;
- understand the value of different knowledge sources and ways of knowing; and
- find, analyze, and synthesize existing data, ideas (e.g., frameworks or models), or methods.

Overview

This chapter introduces the social and behavioral sciences into our understanding of energy systems. Geographer Mike Pasqualetti argues that energy challenges are social problems with the technical component and not the other way around. This is one of the motivations of this textbook: to broaden the horizons of our understandings of energy transitions and its implications. Many energy textbooks cover the fundamentals of energy economics and policy. How tools such as cap and trade, Pigouvian taxes, and tax incentives work are very well described in these books.

Yet, there is also a great deal of scholarship in the social sciences that is less frequently integrated with other sustainability frameworks and assessment tools. This chapter aims to bring social science frameworks such as commodity chain analysis, social impact assessment, the resource curse, geographies of energy, theories of uneven development, and different ways of thinking about justice in energy systems.

3.1 Energy's Impacts on People, and Peoples' Impact on Energy

There is increased interest in understanding and improving the socio-ecological consequences of global production systems. The evolution of society's thinking about production-consumption linkages has shifted over time from an unquestionable good and modernist ideal in some communities. Energy has improved many lives; at the same time, it has had negative impacts on other people and species.

What sets humans apart from other living organisms on our planet is the use of technology. Technology's etymology derives from the Greek root *techne,* which means craft or art, and the root of *–ology* conveys a discipline or field of study. We use technology to grow and prepare food, cloth and house our families and communities, distribute our resources through markets and other financial mechanisms, transport us across the planet and beyond, and to entertain and keep us busy. Technology has improved the living standards and life expectancy of humans, albeit unevenly. But technology has given humans the ability to alter and transform Earth in ways previously unimaginable. We can put humans into space, move mountains into valleys, and bring oil from miles below the sea. Our civilization's rapacious appetite for things and energy has brought considerable disturbance to the Earth's climate and ecosystems, particularly from industrial processes and land-use change for agriculture.

Energy influences many environmental issues. Energy impacts global temperatures and climate variability, both linked to other challenges such as the spread of infectious disease, flooding, desertification, and many others. Energy causes resource depletion as all energy systems are connected to mining and natural resource extraction. For fossil fuels, the fuels are extracted. For renewables, natural resources and land are needed for device construction and generation. Energy causes land-use change, road construction, mountaintop removal for coal production; land-use change also is connected to the climate system. Energy causes soil erosion. The development of corn ethanol in the Midwestern US has led to increased sediment loads in the Mississippi River system. Climate and energy systems also impact the hydrological cycle, which will affect snowpack and water storage, with implications for hydropower and people's livelihoods in India. Energy also impacts biogeochemical cycles as not only carbon but also nitrogen cycles are heavily shaped by the combustion of fossil fuels. Biodiversity consequences are clearly connected to a number of related issues, deforestation, land-use change, and bleaching of coral reefs; even solar power plants can have an impact on biodiversity when not properly or carefully sited. Fisheries and ocean biomass can be affected where power plants take cooling water from estuaries and sloughs, such as we see with several large natural gas and nuclear power plants. Energy systems also lead to waste and materials use, prompting an interest in reusing materials (Sovacool et al. 2020a).

Energy also influences many social issues. Think about how neighborhoods in and around major transportation infrastructure are susceptible to air pollution. Research shows, for example, those living near sources of emissions are more frequently hospitalized for asthma (Maantay 2007).

3.2 Environmental Justice

> "The sad racial history of environmental activism tends to discourage high hopes among racial justice activists. And yet this new wave has the potential to be infinitely more expansive and inclusive than previous eco-upsurges." Community organizer, Van Jones, April 2007

Environmental Justice (EJ) is a term that came into widespread use in the 1980s to describe the uneven distribution of environmental burdens and public health harms (Szasz 1994). The issue has deeper roots to connect it to the civil rights movement. The EJ literature suggests that the locations of incinerators, toxic waste disposal facilities, regional "cancer alleys," and an increased exposure to air pollutants are disproportionately found in low-income or minority communities (Bullard 1996). Therefore, the goal of the EJ activist community is not to redistribute these risk burdens and harms, but rather to resolve them in an appropriate way. To that end, EJ activism looks to breakdown the boundaries of expertise in an effort to democratize decision-making, favor pollution prevention to end of pipe solutions, center justice for workers (Pellow & Park 2002), avoid environmental racism (Cole & Foster 2001), and promote a precautionary approach to environmental policy (Shrader-Frechette 2002).

Definition

Environmental justice—the idea that no social group, communities, or individuals should bear disproportional environmental or pollution burdens.

A new wave of investment is directed toward green energy, clean tech, and nanotechnology. Unlike earlier cycles, this one emerges in the context of increased public attention to climate change, public health risks, and the geopolitical impacts of energy consumption. It also confronts organized social resistance, as grassroots movements, research institutes, and public interest groups call for these investments to be directed toward synergies between climate change, green-collar jobs, industrial ecology, urban renewal, and environmental justice. Will this new wave be "infinitely more expansive and inclusive" (Jones 2007)? Or will these new technologies continue to reproduce structures of inequality in communities, the workplace, and the environment? In India, there is concern that the development of solar farms may be dispossessing rural farmers and agro-pastoralists of their livelihoods (Stock & Birkenholtz 2019). If we are not careful, we could reproduce a decarbonization divide (Sovacool 2020b).

3.3 Energy Poverty

Energy access improves human livelihoods. It may be easy for those with access to modern electricity sources to take for granted that energy is needed for all basic activities, and some households do not have energy to cook, light, heat, or power electrical equipment. Energy poverty is defined as not having access to modern energy sources, or not being able to pay for those energy-related expenses (Robić & Ančić 2018). Over a billion people do not have access to electricity and about three billion people solely depend on fire, and other basic sources of biomass for cooking or heat such as wood, agricultural residues, and dung. Energy poverty occurs in all parts of the world and can have an effect over a person's quality of life in different ways. Access to energy facilitates economic growth, the decrease of unemployment, the promotion of health and education, along with the decrease of poverty. The regions of the world that have the highest levels of energy poverty are sub-Saharan Africa, Latin America, China, and India, and globally, one of five people in the world have no access to electricity. Africa is where energy poverty is perhaps most acute as, excluding Egypt and South Africa, one of five people in Africa have access to electricity. China has seen the most rapid movement out of energy poverty, as they have seen great success with rural electrification.

Two energy indicators are used to determine whether there is energy poverty. The first is access to electricity. Having no electricity constitutes energy poverty in almost all definitions. The second is the type of cooking fuels, for example, relying on wood, charcoal, and/or dung for cooking constitutes poverty. International Energy Agency and other multilateral organizations state that energy poverty is comprised of lack of access to electricity and reliance on traditional biomass fuels for cooking. According to the International Energy Agency (IEA), energy poverty is "inability to cook with modern cooking fuels and the lack of a bare minimum of electric lighting to read or for other household and productive activities at sunset" (IEA 2016). Other groups have tried to quantify these basic human needs met as being on the order of electricity consumption of 50–100 kWh per person per year (think back to the Swiss 2,000-watt society for context) and 50–100 kg of oil equivalent or modern fuel per person per year (United Nations Advisory Group on Energy and Climate Change 2010).

Energy comes in different qualities that can be conceptualized as a ladder, with each step up a higher quality energy type culminating with electricity. The qualities of energy across the spectrum run from simple biomass fuels (dung, crop residues, wood, charcoal) and coal (or soft coke) to liquid and gaseous biofuels and fossil fuels (kerosene, liquefied petroleum gas, ethanol, biodiesel, biogas, and natural gas) to the top tier, electricity (Holdren et al. 2000). More prosperous or wealthy communities use the highest amounts of high-quality energy.

Thinking about Energy Poverty

Consider the following quotes from various practitioners and experts on questions of energy access.

"Energy … transforms the lives of people, communities and nations. No country ever developed without access to energy." —Helen Clark (2011), United Nations Development Program.

"We recognize that, in addition to our separate responsibilities to our individual societies, we have *a collective responsibility to uphold the principles of human dignity, equality, and equity* at the global level. As leaders we have a duty therefore to all the worlds people, *especially the most vulnerable* and, in particular, the children of the world, to whom the future belongs."—United Nations Millennium Declaration.

"Achieving sustainable use requires effective and efficient institutions that can provide the mechanisms through which concepts of freedom, justice, fairness, basic capabilities and equity govern the access to and use of ecosystem services" (Sen 1999).

Communities and families reliant on inferior cooking fuels that are not intentionally using them for other purposes also face the impact of "double exposure" to pollution, globalization, and climate change (O'Brien & Leichenko 2000). According to the World Health Organization (2009), exposures due to the inefficient burning of solid fuels on an open fire or traditional stove indoors creates a dangerous cocktail of not only hundreds of pollutants, primarily carbon monoxide and small particles, but also nitrogen oxides, benzene, butadiene, formaldehyde, polyaromatic hydrocarbons, and many other health-damaging chemicals. Deaths from exposure to indoor air pollution rival the number of global deaths due to tuberculosis and malaria.

Gender implications in situations of energy poverty are particularly acute, particularly in communities where women collect water and energy for household uses. Electricity allows homes to develop water supplies with pumps in rural areas with groundwater availability. Gaining access to energy for a household can immediately reduce the need for family members, notably women, to spend hours per day moving these resources great distances; they can use this time for other activities, such as social activities and entrepreneurship, and for taking rest. Modern energy improves the quality of lighting in homes, thereby improving the well-being of people who might otherwise breathe kerosene or biomass pollution or use inferior lighting such as candles. Lighting neighborhoods and critical infrastructure such as hospitals and police stations also have important roles to improving quality of life of communities in underdeveloped countries. Some argue that modernizing fuels can also reduce deforestation near communities in energy poverty for cooking and heating fuels. LED lantern projects often use purchase or rental frameworks to ensure participant buy-in. These projects suggest significant savings associated with their use compared to kerosene lanterns and candles. Finding solutions to energy poverty are complicated by many entangled issues. Countries may have problems of access, political instability, or be undermined by political corruption.

There are many people working on issues on energy poverty, livelihoods, and climate vulnerability, which are all inextricably linked. Access to energy is an important tool to empower women and enable them to escape some of the exposures and public health impacts from managing households in energy poverty. Many of those without access to electric power have little to no nearby infrastructure, but others who live in the favelas or slum areas outside of large systems have power and fuel infrastructures—transmission lines and pipelines—pass right through them. Solar lanterns are one area where there are demonstrable benefits to those who do not have access to lighting energy, or may be exposed to pollution and high costs of kerosene for lighting. Village electrification has been a major area of energy transitions research since the earliest photovoltaic technologies, as they do not require the same cost-prohibitive level of infrastructure investments. Engineers Without Borders is an organization that has been working on issues related to rural energy development for critical facilities like maternity wards, hospitals, and clinics where electricity provides not only light but refrigeration critical to providing some medicines and vaccines.

Solar dryers in Nigeria are a place where we see how empowering women with technology and tools for success can create entrepreneurial activities in one of the poorest parts of the world. These devices are used in farming communities to dry crops, like rice and peppers, which allow them to last longer on the shelf, which allows growers to sell a larger portion of their harvest. New business models that allow poor farmers to rent equipment and pay with the revenues of crop sales will be essential to ensuring that groups that lack capital have access to what they need to make projects work. Solar-powered micro-grids built off-grid are a solution that is attracting investment in pilot projects. But these are still in the early years of development.

3.4 Resource Curse

An important topic in the study of natural resource extraction is the resource curse. Some scholars refer to this as Dutch disease. The general proposition is that some communities become over-reliant on one commodity either for employment or for government revenues. One classic case is Nigeria, where the country has tremendous petroleum wealth that is lost through circuits of poor policy decisions, corruption, or just inequality. Geographer Michael Watts has posited something more complex than the resource curse per se that involves flows of power away from marginalized communities toward, in some cases, the military industrial complex of Nigeria. Benjamin Sovacool (2014a) examined the question of resource curse for communities with shale gas.

Definition

Resource curse—the apparent contradiction that some communities become over-reliant on an extractive natural resource and that social systems are unable to ensure the community benefits from these riches.

3.5 Behavior and Energy

One major contributor of carbon dioxide emissions is what people do in their everyday lives and in their own households. Social sciences focused on energy and behavior not only has sought to understand the potential to change energy use patterns but also finds that the approach will face major obstacles (Stern et al. 2016). There is a widely cited quote by Amory Lovins that says, "people do not care about where their energy comes from, all people want is warm showers and cold beer." Household energy use is approximately 8% of global emissions, second only to China in overall emissions by country (Dietz et al. 2009). The practices that are categorized as behavioral practices range from the implementation of sustainable measures in heating ventilation and cooling (HVAC) in homes to shifts in activities. Research shows that 7% of energy use could be reduced from behavioral changes alone (Dietz et al. 2012).

Public perceptions of the adoption of energy conservation measures are categorized in a wide spectrum from little to no information of how households can practice conservation, sustainable building practices, and low-carbon HVAC implemented in new homes (Attari et al. 2010). Increased public outreach is necessary for state and government agencies to maximize the incentive programs already in place that target the reduction of greenhouse gases in businesses and residential properties. The issue with many incentive programs is that customers are overloaded with information, and correctly applying for the incentive can be complicated. Informational programs that provide descriptions of content and workshops make it easily accessible for the public to gain knowledge on programs that advocate for the conservation of our resources while also providing the public with the option to get further knowledge in other practices they can adopt as well. Education maximizes the way we advocate our incentive programs and our conservation tools.

Some social scientists have explained energy behavioral shifts in the context of altruism. In one study, households were provided with feedback on their energy cost savings compared with the most efficient 10% of their neighbors. A second group of households received feedback about energy consumption, and instead of telling these customers how much they saved, the savings were expressed as avoided air pollution. In other words, one group received information about efficiency and their personal savings, while the other got information about the collective good of reduced air pollution. The findings are interesting because they show a stronger behavioral response to avoiding air pollution than personal financial savings, which is contrary to how many energy efficiency programs are marketed. Electric utilities market energy efficiency savings in terms of dollars saved, but it turns out that people may be more responsive to the altruistic goal of avoiding something that contributes negatively to public health.

Another area of behavioral research around photovoltaic adoption is the residential sector. These are homes that would be prosumers—a term that connotes they produce and consume. Economists would argue that prices are logical drivers. So if a photovoltaic installer proposes a project that is cheaper than an electricity plan from the electric utility, most electric consumers have the incentive to go solar. But there are other factors at work that could be sociological or psychological such as the "neighborhood effect"—where the peer effects of neighbors could help hasten adoption of rooftop residential solar energy as people start to know solar adopters (Graziano & Gillingham 2014).

William Stanley Jevons was an economist who asked how technological advances and efficiencies might have unforeseen negative effects. In his 1865 book, *The Coal Question*, Jevons (1906) explained the mechanism whereby energy efficiency improvements lead to increased energy consumption. "If the quantity of coal used in a blast-furnace, for instance, be diminished in comparison with the yield, the profits of the trade will increase, new capital will be attracted, the price of pig-iron will fall, but the demand for it increase; and eventually the greater number of furnaces will more than make up for the diminished consumption of each" (Jevons 1906). This is known as the Jevons paradox, and it is one way the rebound effect can occur.

Definition

Rebound effect—this idea asserts that gains in energy efficiency are not always realized because of other systematic effects. For example, savings from fuel economy improvements can lead to increased driving. Or, more efficient heating could lead people to keep their homes warmer in the winter. A similar concept called the Khazzoom-Brookes postulate.

There are multiple interactions that could cause the rebound effect—some direct, others indirect. For households, direct rebound effects derive from increased consumption of energy services, such as heating or lighting, where effective price has fallen as a result of improved energy efficiency. For example, the replacement of traditional lightbulbs with compact fluorescents will make lighting cheaper, so people may choose to use higher levels of illumination or not switch lights off in unoccupied rooms. There are a range of studies of rebound, with general findings of an effect of 10–20% depending on the specific technology or intervention. The Khazzoom-Brookes postulate asserts that economy-wide increases in energy efficiency counterintuitively make the overall energy use increase. This postulate, or hypothesis, does not directly link the "rebound" to behaviors or responses to economic signals. Reasons hypothesized are related to the cost of energy relative to the other energy-using activities that amplify energy use.

The backfire effect occurs when an energy efficiency intervention results in a higher amount of GHG emissions or energy use. Space heating seems to be the second biggest contributor to the rebound effect, and could also backfire, because as heating and air-conditioning technologies become more efficient, the public tends to use the appliances more. Heating and cooling requirements are different in different locales, compare for example California to the East Coast in winter.

Air-conditioning, on the other hand, is used far more often in California because of the heat, particularly away from the coast. Heating a room requires a lot of energy, but cooling a room, on the other hand, is far less energy-consuming than heating. Cheaper air-conditioning means that the public could use it far more often, which would be an example of this rebound effect. According to a study done by the European Commission, air-conditioning alone has a possible rebound effect up to 50%, with an average of 25% (European Commission 2011).

Space heating and cooling are the biggest contributors to the energy use of residential buildings. As more use of smart meters coupled with economic incentives for displacing or reducing energy use could play a major role in future grid management. Consumers may even be responsive to better understandings of their energy use patterns. A study of a dozen separate trials from the US, Canada, Australia, and Japan used direct feedback via an electronic in-house display and found the device results in demand reductions, ranging from 3% to 13%, or on average 7% (Carroll et al. 2014).

The role of human behavior and decision-making is an important area of research needed to better understand both energy efficiency and renewable adoption. The "energy efficiency gap"—the gap between those who actually make the investments and those who have the opportunity to recoup in short term, but do not—is an important area of research. The fact that this gap persists suggests that energy efficiency investments with short economic pay-back periods are underutilized, and perhaps there is a need for more feedback and reminders related to how energy efficiency pays off in terms of energy savings (Gillingham & Palmer 2014). Energy efficiency is important because while some decarbonization efforts will cost the public or ratepayers money, these energy efficiency ones usually pay off in a few short years.

3.6 Theories of Social Change: Ecological Modernization and Social Movements

Several social science disciplines explore theories of social change, including sociology, geography, anthropology, economics, and political science. In sociology, it is social movement theory that most closely works on these questions related to strategy as well as descriptive accounts of social change. Since energy transitions are the topic of this book, this body of social theory is particularly relevant. Energy transition groups can be classified as those advocating transformation of a particular energy system or socio-technical change and those aiming to achieve some broader social change (Hess 2018). Some of the key questions for social movements looking to reshape the future of energy are discussed in the following paragraphs.

A common question that often enters into questions of social change is whether social movement organizations should choose collaborative versus adversarial strategies. While protest often plays a significant part in explaining social movements success, other groups derive their success by being narrowly focused on specific policy-level interventions where industry collaboration is required. While on the surface it seems that both strategies must guide social movements, the discursive use of collaborative proposals that promote "ecological modernization" and "industrial ecology" make it difficult to identify what the outcomes of collaborative strategies actually are (Spaargaren & Mol 1992). In this context, the win-win solutions promoted by emphasizing collaboration can be disempowering to social movement goals in a larger context. Adversarial approaches include protests, lawsuits, and boycotts. They tend to rely on tactics that are disruptive. A famous thesis on "poor people's movements" found that disruptive protest often had important ramifications for the social policies (labor organizing and civil rights) of the welfare state (Piven & Cloward 1979).

Questions social movement theorists ask include whether external or internal variables affect a social movement's ability to make change. Do movement-controlled facets or some aspect of the movement's environment determine outcomes? These perspectives consummate the differing perspectives of resource mobilization, political process, and new social movement explanations of success. Most theorists draw on various elements of these perspectives that best capture the dynamics of their case studies.

Resource mobilization explanations suggest that social movements have the ability to effect social change if they can mobilize the appropriate resources. These studies look at the reliance on money, leadership, political allies, and organizational form, focusing on internal characteristics of social movements. This framework relies on theories about collective action and uses notion of rational behavior to explain why people participate in social movements. Recent studies raise questions about how narratives and framing are mobilized as resources. For example, discursive representations of the appropriate scale for policy intervention seek to delegitimize grievances and can neutralize mobilization.

Other studies focus on the ways that mobilization is geographically constituted through space, place, and scale, recognizing that resource mobilization is place-based, is place-structured, or employs a spatial strategy. For example, a study of antinuclear activism in the Boston area through the 1960s and 1980s found that place-specific conditions—class structure, histories of activism, and educational levels—shaped peace campaigns differently in each locale. In some places, appeals of antinuclear activism began to waver as it became clear that it undermined the material interests of those working in the Boston area arms industry.

Political process explanations relate the ways that institutional structures are open or closed to political protest (McAdam 1982). They privilege both the political context and the reconfiguration of power relations of specific situations they call opportunity structures. These opportunity structures are temporally and spatially both specific and uneven; opportunity structures open and close depending on the circumstance. One comparative study of antinuclear movements in four nations found that their success depended on whether political opportunity structures let actors into the policy process (Kitschelt 1986). In studies of other emerging technologies, similar patterns exist. Genetic engineering could be important to future biofuel technologies. In one study of the anti-genetic engineering movement, they found that industry opportunity structures allowed activists to exploit consumer anxieties for genetically engineered organisms, directly engaging with the financial decisions made by private corporations (Schurman 2004).

Geographers borrow from sociologists in developing more spatially based theories of social change. New opportunity structures are opened or closed by "jumping scale," a notion used in the geography literature to describe a process where actors move to new scalar jurisdictions to seek out advantageous power asymmetries (Smith 2010; Swyngedouw 1997). These power asymmetries are not static but constantly shifting. The Christian Right movement, for example, to gain political power mobilized support through local institutions like school boards and churches until favorable opportunities appeared after many years of social and cultural changes (Diamond 1998). Questions about social and cultural changes suggest broader themes about identity formation, the topic of the new social movements' explanations. New social movements are those movements that are not identifiable by class-based concerns, but that are characterized by the defense of identities, ideas, and practices against the intrusion of the state and economy. These actors occupy both class and cultural positions—what Jurgen Habermas (1981) called the system and the lifeworld. Mobilizing new social movements requires understanding the material places and relationships being defended, the ways actors conceptualize grievances and form identities, and the articulation of these in the spaces and places of interaction (Miller 2000).

In sociology, this idea of ecological modernization embodies this notion that environmental problems are at root design and policy problems (Berger et al. 2001). Minimizing environmental pollution or redirecting environmental contaminants toward facilities where they can be utilized as feedstocks can avert many of the consequences of the treadmill of production, according to key tenets of environmental sociology.

Yet, many argue that it will be green technology that steers human civilization away from planetary catastrophes and the planet, as it replaces conventional technologies with more environmentally benign ones. As the human-environment relationship evolves, it is possible that technologies can be deployed to make that relationship more sustainable. Renewable energy promises to lessen our impacts to the extent to which it can be deployed. More efficient resource utilization through phenomena like industrial symbiosis and cradle-to-cradle design will lead to materials' reuse and recovery, and it will lower rates of raw material acquisition. Smart grids, for example, are designed to utilize energy more efficiently and encourage more energy conservation. This technology-substitution emphasis is usually accompanied with the argument that green markets will drive changes and innovation as the environmental externalities created by the economy are internalized, as market prices reflect the environmental costs of doing business. There is obviously some truth to the claim generally. But in the real world, there are caveats abound, so it is still best to treat the market efficiency claim as a hypothesis rather than an absolute fact.

3.7 Political Ecology

Political ecology draws on concepts that range from qualitative to mixed methods such as global production networks, commodity chains, and other conceptual apparatuses to link together natural resource extraction to the goods and services delivered to human civilization. Political ecology is an interdisciplinary field that cuts across environmental history, cultural anthropology, environmental economics, politics, and even some of the natural sciences such as ecology, making it uniquely situated to understand energy transitions. The subfield's main thrust has been two-fold. First, as a research agenda, it has interrogated assumptions about what causes environmental change that it sees as problematic because they de-emphasize the political and social causes of environmental problems; these problematic characterizations of environmental change are apolitical. These apolitical ecologies, as Geographer Paul Robbins calls them, tend to naturalize environmental problems or make them seem as though they are inevitable (Robbins 2004). These challenges apply across the board to different environmental issues. One study of discourses around biodiversity, desertification, climate change, and deforestation found that the way that these environmental problems were framed in terms of both causes and solutions did not match what they found when they explored these problems in specific locations (Adger et al. 2001). This attentiveness to how problems and solutions are framed discursively and how they shape and are shaped by institutional responses and power dynamics is a strength of political ecology approaches.

Environmental politics often involves struggle over problem definition, where actors try to make others see problems according to their views, and position other actors in specific ways. Narratives of environmental problems typically have elements of causality, responsibility, victimization, and intentionality. The major focus of discourse analysis is to move beyond simple representations of conflict over single issues to understanding the multi-faceted and complicated roots of some environmental problems. Discourse here is meant as some shared meaning of some phenomenon, such as desertification (Adger et al. 2001). Arguments about regulatory capture or conflicts of interest are based on statements that do not lend well to actual depictions of causality. It does little good empirically to show that

Monsanto influences regulatory science by way of its complex personal connections (the revolving door) or entrenched interests (regulatory capture). Instead, discourse analytic approaches seek to link interests to outcomes in a more causal way, drawing on other facets of social theory to explain why some have power over others and in what contexts.

In political ecology, discourse analysis is used to identify how environmental problems become naturalized or deemed inevitable. Discourse can be taken to mean "a set of linguistic practices and rhetorical strategies embedded in a network of social relations" (Litfin 1994). A discourse can be a set of claims that substantiate and legitimate an ideology. Political scientist John Dryzek (1997) describes discourse as "a shared way of apprehending the world. Embedded in language, it enables those who subscribe to it to interpret bits of information and put them together in coherent studies of accounts. Each discourse rests on assumptions, judgements, and contentions that provide the basic terms for analysis, debates, agreements and disagreements, in the environmental area no less than elsewhere" (Dryzek 1997).

Many political ecology approaches look at how the meanings of environmental problems are contested and seek to reconcile the various ways of seeing problems and explain why some interpretations of environmental problems are hegemonic, influential, or discredited. Key topics in this field are vulnerability, resilience, colonialism, environmental racism, and natural resource degradation and management.

In the area of energy and the environment, many researchers study vulnerability to natural hazards or climate-changed enhanced storms and weather. For example, geographer Susan Cutter points out that hurricanes in poor areas have become part of a vicious cycle of vulnerability (Cutter et al. 2008).

A Rhetorical Model for Environmental Discourse and Its Political Discourse

Ethos: relies on the reputation of the speaker and the appeal to character; regulatory discourse; *Example*: scientists agree that the products of biotechnology present no new risks.

Pathos: relies on how emotions are evoked; poetic discourse; nature as spirit; evokes values; *Example*: anti-genetic engineering activists are causing people to starve in Africa.

Logos: relies on how persuasive a case is made; appeals to fact and reason; *Example*: food security depends on advances in plant breeding. Biotechnology is the most important new plant breeding tool. We need to utilize biotechnology to improve food security.

These three categories are important because they tell us something about why some appeals are more successful than others. The rhetor chooses particular discourses because the rhetor believes that it helps them persuade. For example, the rhetor knows that the people can be persuaded by invoking questions about hunger and starvation because the rhetor knows that such arguments are representative of some of the values held by society. Likewise, an appeal to logos shows that the rhetor must be able to draw out reasonable and logical arguments. Finally, because much of persuasion is built upon trust, the rhetor must appeal to ethos.

Some discourses are more persuasive than others, but not all the time. The point is to learn something about the broader social processes affecting the problems in question. The job of the rhetorical analysis here is to understand/reveal the underlying strategies and assumptions. How does the audience make sense of this discourse? What might they be taking for granted?

The political-ecological approach sees the mycelia of actors and relationships that comprise commodity assemblages as an emergent phenomenon sustained by networks of materials, knowledge, finance, politics, and culture. Who takes value from where along the commodity chain and its associated chain of accumulation? Developing causal models is essential for environmental problems. Without knowing what contributes to what and without knowing which contexts matter and meaning of these finding for praxis is central to the debates about explanatory chains (Blaikie & Brookfield 1997).

Questions of access and control of natural resources are at the core of many research questions in political ecology. This research community studies the politics of natural resources with a deep understanding of the socio-ecological actors that shape interconnected systems of production and consumption with an emphasis on power and justice (Watts 2000). Political ecologists who draw on commodity chain approaches often ascribe agency to nature-human relationships, allowing non-human objects and ideas to shape as much as be shaped by socio-ecological processes (Bennett 2009; Lockie & Kitto 2000). Much of this research links developed nations' consumption patterns to land degradation in developing countries. Many commodity chains have resulted in environmental pollution or poor treatment of workers, that is in part due to the distances over which investment and consumption decisions are made, which mask their consequences. Consequently, sustainability is an important theme to correct some negative consequences of accountability gaps created at a distance (Caniato et al. 2012). Some of these forays into the sustainability of GPNs stem from a growing interest in corporate social responsibility (Bryant & Goodman 2004).

Karl Marx argued that technological development in capitalist society is exploitative and alienating. But he remains committed to the idea that working-class struggles can regain control of technological development to suit the purpose of masses. Scholars suggest that there are fundamental contradictions between capitalist development, which some argue hits upon barriers posed by nature that capital is unable to overcome (Bellamy Foster 2008).

More contemporary scholars reject the teleology of Marx and destabilize the notion that technologies are motivated by, and behave in, intended ways. Actor network theory is a framework that emphasizes contingency and unintended consequences in technological design and deployment.

Commodity chain analysis also poses interesting questions about how power flows along the commodity chain. For example, there is a distinction between producer-driven and supplier-driven or buyer-driven by year-driven commodity chains. So the ability of firms to influence up the supply chain depends on who holds power in the commodity chain. Big purchasers like Walmart might be able to influence every producer that hopes to sell products in their stores. On the other hand, big chemical companies like Dupont that sell inputs to the photovoltaic industry might dictate the production systems and items that the solar industry purchases. Power can flow either up or down the supply chain, depending on the industry structure and position of actors involved.

Others questioned whether contradictions in capitalism make meaningful environmental reforms impossible (O'Connor 1994). Sustainability claims may be undermined by its coexistence with mass consumption (Luke 2005). A few compelling critiques of CSR as greenwashing reinforce the concern that CSR takes attention away from or delays regulatory interventions.

Concepts like commodity chain analysis and life-cycle assessment as analogous tools that trace out production systems in very different ways. One focuses on materials and energy, while the other at social relations. Bringing these ideas together might be a fruitful exercise in studies of energy transitions because energy systems too have these complex linkages that are global in nature, structured by discourse, and shaped by power relations and asymmetries.

3.8 Global Production Networks

The GPN framework, advanced by researchers in economic geography, political ecology, and sociology, is used to answer various research questions from understanding colonialism, patterns of economic development, and global governance to the socio-ecological transformation of natural resources into commodities and implications for labor. The basic concept explains how production systems are interconnected to produce global commodities. These range from relatively simple raw materials to complex products like computers.

The trend in our global economy is toward increased interconnected economic and cultural relationships. A global commodity network is nothing new though. Many commodities and products have been produced through global systems of production for centuries, if not longer. The distinction is that global systems of production are more widespread and integrated into major economic flows, as technology and other social forces of globalization are reshaping finance, capital flows, the ways global products are made, and their geographical composition. Containerization, information systems, telecommunications, and cheap energy have reshaped how production systems can "act at a distance" more than ever before. At the same time, multinational actors are more powerful than ever as well. Some hold that these actors are also less unaccountable as their power might be seen to exceed the jurisdiction of nation-states. So an objective of GPN research is to illustrate the political economic forces that shape GPNs, allowing for a deeper understanding of how commodities are constructed and their consequences for people and the environment.

3

Global Production Networks, Value Chains, Filieres, Circuits, and Commodity Chains

Global production networks	In research that uses the concept of GPNs, there is a tendency to focus on the behavior of multinational actors and institutions.
Global value chains	The concept of global value chains aims to capture the activities that give rise to global production systems (Gereffi et al. 2005). Firms construct value over space through sourcing and contracting arrangements, and this approach aims to understand how these activities are organized and governed. Value is added across the supply chain as materials move from raw materials to finished products (Gereffi et al. 2001)
Filieres	A French concept that seeks to explore the chain of activities related to the production of raw materials into final export products (Mather 1999). Research on Filieres usually follow the commodity beyond its useful life, as opposed to other analyses, which may stop at the factory gate/site of production.
Global commodity chains	There are several clusters of research (sometimes overlapping) that describe their unit of analysis as the global commodity chain. These include some sociological work that comes from world systems theory (Hopkins and Wallerstein 1986) and work in political ecology (Robbins 2011).
Commodity circuits	Studies of commodity cultures prefer the concept of circuits to chains because the metaphor does not imply a start and end point (Cook & Crang 1996).

3.9 Social Acceptance of Energy Systems

The social acceptance of energy systems from infrastructure to the adoption of different modes and sources of energy promises to be an important area of study. The "social gap" in renewable energy is the space between those who support renewable energy generally to those who support particular projects (Bell et al. 2005). This research finds several explanations for public support for projects, including the structure of public participation, the acceptability of major impacts in the decision-making context, the degree of self-interested NIMBYism, major differences in values and beliefs (Bidwell 2013), the degree of stakeholder collaboration (Phadke 2011), and perceived or real inequitable power relations (Bell et al. 2005). Better understandings of public attitudes (Walker & Cass 2007) and socio-technical imaginaries toward energy futures may hasten deployment, create more cooperation, and improve socio-ecological outcomes (Jasanoff & Kim 2009).

There is also a dearth of research on environmental justice impacts from siting utility-scale solar energy facilities. Utility-scale solar power plants do not have the same externalities as fossil fuels used for electricity generation. Yet there are impacts from these facilities, including land use change and the nuisance of noise and dust. There have also been several lawsuits in Western US directed at solar developers by Native Americans over burial grounds and sacred sites (Mulvaney 2019). Collaborative planning between resource agencies and tribes could help foster solar energy deployment on appropriate lands and mitigate any environmental justice issues. The concept of social barriers to changing energy landscapes helps us collect and evaluate concerns that publics might have with the development of renewables (Pasqualetti 2011). For example, one explanation is that there is a democratic deficit with decision-making around energy deployment. This research shows how the "decide-announce-defend" approach to project development makes any social resistance difficult to overcome. More participatory approaches where communities shape project outcomes can hasten clean energy deployment.

Several studies of the social gap in support of wind farms suggest concerns about loss of property value. The first round of wind turbine installations can lower home values because of the visual impact, but over time, these prices recover, as new wind farms become part of a landscape or location; so research even shows wind farms can have a positive effect on the housing price (Pasqualetti 2001).

Avian (bird) collisions with wind turbines is an increasing concern for large birds of prey like raptors (Liechti et al. 2013). Bird collisions with wind turbine is an ecological concern because of the growth in wind farms across the landscape and our lack of understanding of avian perceptions (Bernardino et al. 2018). In order to reduce bird collisions, scientists suggest three strategies: "(1) further development of the methodologies used to predict impacts when planning a new facility; (2) assessment of the effectiveness of existing minimization techniques; (3) identification of new migration approaches." The main causes of bird fatalities are either species-specific, site-specific, and wind farm-specific. Species-specific evaluates certain bird species that are more likely to fly through a wind turbine and power lines. Large birds would often use thermal updraft to reach very high altitudes to fly long distances. A thermal updraft is a mass of hot, rising air that birds can often rise in just by holding out their wings. The susceptibility of wind farms to high rates of collisions is heavily dependent on landscape features for wind turbines and usually very site-specific.

Ridges, steep slopes, and valleys are mainly used for birds, but when constructing wind turbines in their habitat, it results in biodiversity loss. Many birds have certain flight paths, and having a wind turbine in the middle of their flight path presents a risk. Wind farm-specific are turbine features that play an important role in bird collision risk. There are strategies that can direct birds to choose a different flight path to avoid, minimize, and compensate collisions. It is better to plan early before construction on wind turbines to see where the flight path is for bird species. Minimize the impact magnitudes through the implementations of single or multiple measures of the wind turbines. Critical to the conservation of bird species of concern may be to empower on-site biologists to depower turbines as a means to reduce bird collisions. Doing so may require special adjustments to turbines as depowering can potentially harm the turbines. Another mitigation measure to use is increase visibility of wind turbines. If the wind turbines are more visible, it may help reduce avian collisions.

3

Assignment 3.1 Exploring Issues Across the Social Gap in Renewable Energy Deployment

In this assignment, we develop an understanding of real-world controversies over siting renewable energy projects. The case we explore is about a contentious wind farm in California's Colorado Desert, near the US-Mexico border. It would deliver electricity to metropolitan San Diego via the approved, but also controversial, Sunrise Powerlink, a large transmission project that crosses Anza Borrego State Park. Some stakeholders involved in the controversy see these desert landscapes as ideal sites to enlist in the fight to decarbonize electricity. Others have come to understand the wind farm as a severe ecological impact on a site rich in biodiversity, wildlife, and cultural resources. This seeming paradox is called the "social gap" in renewable energy deployment—that there is strong, consistent support for renewable energy, but widespread local opposition to siting particular projects. We will recreate this wind farm controversy to understand the multiple stakeholder perspectives, and propose solutions.

The key challenges to this project include the project's location on public lands of interest to Native American tribes, environmental organizations, and other stakeholders in the area. The project occurs in a region with very high unemployment rates and could provide more economic opportunity in the area. The use of public lands for energy development poses some intractable problems for both cultural resources—which include living organisms like the flat-tailed horned lizard, core to the creation story of the Quechen Tribe—and ecosystems because many public lands are in conservation by default. The most critical issue as in many wind farm cases is the importance of mitigating and avoiding avian collisions. When the project was built, the wind farm operator agreed to build an avian monitoring tower and radar to help avoid golden eagle deaths in 2011. However, in 2015, personal communications between the Fish and Wildlife Service and the Bureau of Land Management (BLM) suggested that this tower and radar system was not working. This case points to the challenges of using public lands that are of high habitat quality for energy development, when more degraded sites on private lands are available.

Learning Objectives

- Advance energy literacy by improving student understanding of where energy comes from.
- Develop skills to communicate about energy issues in meaningful ways.
- Describe and evaluate stakeholder views offer critiques various positions in an environmental debate.
- Improve understandings of socio-environmental consequences of wind power.

- Assess the credibility of information about wind farms and their impact on socio-environmental systems at and across multiple temporal and spatial scales.
- Learn about the public comment process during the environmental impact statements required under the National Environmental Protection Act/California Environmental Quality Act.
- Read and analyze public comments as existing data to understand how stakeholders value and construct place and how subjectivity informs public acceptance of wind projects.
- Reflect on opportunities to improve public participation.
- Evaluate mitigation options to lessen the impact of the wind farm on wildlife and biodiversity.

Preparation
Leave yourself 2–3 hours to read and take notes on the following articles. The first short online book chapters introduce key principles of wind power. The three academic research articles describe the socio-environmental impacts of wind power and some of the explanations for social acceptance or resistance toward wind power. While reading the articles in the chapter references (Bell et al. 2005, MacKay 2009, Tabassum-Abbasi et al. 2014, Van der Horst 2007), make note of the following: (1) negative and positive impacts from wind power, (2) explanations for the "social gap" in wind energy, and (3) whether NIMBY is a good explanation for social acceptance of renewable energy.

Lecture and Discussion
In class, we will have a lecture of approximately 45 minutes where you will be introduced to fundamental concepts for understanding issues related to wind power. Topics in this lecture include: sources of the energy in wind power, power potential from wind, Betz' law, current extent of global wind power installations, geographies of wind, technical challenges for more widespread wind power integration, and socio-environmental dimensions of siting wind power.

Two-Hour Homework
Now that you have a deeper understanding of the challenges related to siting wind farms, we shall review a case study of the Ocotillo Express Wind Energy Facility (OWEF). There are three parts to your preparation, which should take about two hours. For class, you will bring a printed copy of the work for Part 2 and Part 3.

Part 1. Read an overview of the issues related to constructing the wind farm (OWEF). The Final Environmental Impact Statement can be found at this link ▶ http://www.icpds.com/?pid=2843

Read these sections: Dear Reader, Abstract, and Introduction. ▶ ftp://ftp.co.imperial.ca.us/icpds/eir/ocotillo-express/final/05introduction.pdf. These materials are all also archived at www.dustinmulvaney.com/OWEF and www.archive.org.

Part 2. Draw a concept map that shows the issues and stakeholders involved in the OWEF controversy. Go to ▶ http://www.mentalmodeler.org/online/ and click "add component." Type "Ocotillo Express Wind Energy Facility (OWEF)" in the box and drag it to the center of the screen. List each of the following stakeholders as additional components and place them around the

OWEF box: climate-focused environmental groups, wildlife-focused environmental groups, Native American tribes, chamber of commerce, opposed local citizens, local citizens in favor, the wind farm developer, the BLM, and any other actors you come across in your preparation. Click the OWEF box and hold down the arrow below the box to draw arrows from the OWEF box to each of the stakeholders you've listed. Once the arrows are drawn, click on the box in the center of each arrow and click whether each arrow represents a negative or positive view of the wind farm. You can distinguish whether these views are more or less strongly held by selecting multiple plus or minus signs, which will result in thicker or thinner lines.

Next, based on your understandings of perceptions of wind farms from the readings, list some impacts that each stakeholder might be concerned about. For example, the local chamber of commerce would be interested in jobs, tax revenue, and increased local business, while the Native Americans would be concerned about cultural resources and sacred histories. This concept map should contain all the stakeholders and the issues that may be raised by each (Fig. 3.1).

PRINT, make a PDF, or take a screenshot of this draft concept map for class meeting #2. You will turn this in with your final work product, but feel free to mark it up with notes from class. You will revisit this diagram in the last steps of this assignment.

Part 3. Each student is assigned one of the groups listed in bold below. To prepare for class meeting #2, write a 200–300-word summary that explains the position of the groups and institutions that you are assigned to represent. The wind farm developers should prepare a summary focused on overview of the project (developer and BLM district office) and agency's role (BLM). The other stakeholder summaries will be read as two-minute public comments. You will share this summary with your peers in small groups during class time. Stakeholder public comments can be found in the Imperial County EIS website: ► http://www.icpds.com/?pid=2843

Three PDFs labeled Comment Letters 1, 2, and 3 are at the bottom lower right of the web page in Appendix O. The course instructor should print out or collect the PDFs for several of these and pass out to the participants.

Fig. 3.1 Divided views on the California wind farm

See below for instructions for where to find information and background material. *If any websites are missing, enter them into this site:* ▶ https://archive.org/

Wind farm developers—Prepare a project overview and project benefits.

Pattern Energy ▶ http://www.ocotillowind.com Also see: Ocotillo Wind Energy Facility (OWEF). 2012. Eagle Conservation Plan.

Final EIS/EIR. Read: Fig. 1, Fig. 2, Fig. 3, Section 1.3, Table 2, Fig. 6, Fig. 8, Fig. 9, Sect. 5.1,

▶ ftp://ftp.co.imperial.ca.us/icpds/eir/ocotillo-express/final/appendices/20appl9-eagle-conservation.pdf

Bureau of Land Management, El Centro District Office—focus on BLM's role in the project.

▶ http://www.blm.gov/ca/st/en/fo/elcentro/nepa/ocotillo_express_wind.html read: the fact sheet, abstract, and webpage.

Bureau of Land Management, Wind PEIS Office

▶ http://windeis.anl.gov/ read: What is an EIS? What is a Programmatic EIS? ▶ http://windeis.anl.gov/faq/index.cfm

Fish and Wildlife Service

Read pages 1–12 and 51–53 Biological Opinion on: ▶ http://www.blm.gov/ca/st/en/fo/elcentro/nepa/ocotillo_express_wind.html

Imperial County Board of Chamber of Commerce

See "comment letter 1" file on Imperial site mentioned earlier, starting at page 322

Kumeyaay Tribe + Cocopah Tribe + Quechen Tribe

See "comment letter 1" file on Imperial site mentioned earlier, first few dozen pages are relevant. Also see statements on the Internet such as ▶ http://www.kumeyaay.com/all-news/2908-eight-tribal-nations-mourn-losses-at-ocotillo-wind-site.html ▶ http://www.kpbs.org/news/2012/mar/19/tribes-fight-green-energy-wind-project-desert/

▶ http://eastcountymagazine.org/node/9104

▶ http://www.kcet.org/news/rewire/wind/imperial-county-protest-halts-work-on-wind-project.html

Defenders of Wildlife + Natural Resources Defense Council + Wilderness Society

▶ http://wilderness.org/sites/default/files/legacy/Ocotillo-Express-Scoping-Comments.pdf

Sierra Club

See "comment letter 1" file on Imperial site above, starting at page 309. Also see

▶ http://www.eastcountymagazine.org/suit-filed-halt-ocotillo-wind-coalition-holds-protests-san-diego-and-el-centro

Save the Eagles International

Ocotillo Wind Energy Facility (OWEF). 2012. Eagle Conservation Plan. Final EIS/EIR.

Read: Fig. 1, Fig. 2, Fig. 3, Section 1.3, Table 2, Fig. 6, Fig. 8, Fig. 9, Sect. 5.1,

▶ ftp://ftp.co.imperial.ca.us/icpds/eir/ocotillo-express/final/appendices/20appl9-eagle-conservation.pdf

California State Parks Foundation

See "comment letter 1" file on Imperial site above, starting at page 323

Desert Protective Council

See "comment letter 1" file on Imperial site above, starting at page 326. Also see

3

▶ http://www.eastcountymagazine.org/suit-filed-halt-ocotillo-wind-coalition-holds-protests-san-diego-and-el-centro

Center for Biological Diversity

See "comment letter 1" file on Imperial site mentioned earlier, starting at page 360)

Basin and Range Watch

▶ http://www.basinandrangewatch.org/OcotilloWind.html

Concerned Local Citizens

See "comment letter 1" file on Imperial site mentioned earlier; starting at page 429, there are comments from public citizens; comment letter 2 has many more. One student should read several perspectives to get a range of local views.

In-Class Group Activity

You will work in small groups to summarize the details of particular views and perspectives on the proposed wind farm project. First, we will gather in small groups to go over the highlights of issues raised by their particular stakeholders. Groups that are most appropriate to put together are in a list below. We will have 20 minutes to discuss in groups. Each student should take turns describing their stakeholders' reasons for involvement based on the summary they prepared. Where more than one student is assigned to a stakeholder, students should decide who will present the perspective to the rest of the class in the public comment activity and who will volunteer to write the notes on the white/smart/chalk board.

Group A: *Wind farm developers + Bureau of Land Management (BLM), El Centro District Office*

Group B: *Bureau of Land Management (BLM), Wind PEIS Office + Fish & Wildlife Service*

Group C: *Kumeyaay Tribe + Cocopah Tribe + Quechen Tribe*

Group D: *Defenders of Wildlife + Natural Resources Defense Council + Wilderness Society + Sierra Club*

Group E: *Imperial County Board of Chamber of Commerce*

Group F: *California State Parks Foundation + Desert Protective Council + Center for Biological Diversity*

Group G: *Basin and Range Watch + Save the Eagles International*

Group H: *Concerned local citizens*

Once your group has discussed the views of stakeholders represented, we will reconvene the entire class. For the next 40 minutes to an hour, the groups will present stakeholder perspectives. To mimic a public meeting, we will begin with the BLM El Centro District Office who should provide an overview of the need for the meeting and some of the key issues BLM will evaluate. Next, the wind farm developers should present the wind farm proposal and the benefits of the project. After the introduction to the BLM and project, the public comment period should include each stakeholder perspective. One student per stakeholder group will sign up for a two-minute speaking slot with remarks that are prepared beforehand (or two students can speak for one-minute each). Several volunteer notetakers will summarize on a white/chalk/smart board. These notes will be useful for your final work product.

Opening Statements

A public meeting is usually opened with a representative of the US Bureau of Land Management (BLM), district office. In this case study, that is the El Centro District Office. The opening should describe the project and summarize the major impacts that have been identified to date, as well as outline the rest of the process. Sometimes, the project developer (wind farm developer in this case) or lead environmental impact assessment consultant will speak to the project or be available for question and answer after the public comment period.

Public Comments (Two Minutes Each)

1. Bureau of Land Management (BLM), Wind PEIS Office
2. Fish & Wildlife Service
3. Kumeyaay Tribe
4. Cocopah Tribe
5. Quechen Tribe
6. Defenders of Wildlife + Natural Resources Defense Council + Wilderness Society + Sierra Club
7. Imperial County Board of Chamber of Commerce
8. California State Parks Foundation + Desert Protective Council + Center for Biological Diversity
9. Basin and Range Watch
10. Save the Eagles International
11. Concerned local citizens
12. Close of the public comment period

Class Discussion

Once the public comment period has closed, we will reflect on what we can conclude about the controversy. How well do some of the explanations for the social gap in renewable energy help us understand the roots of this particular controversy? Which explanations seem to be less relevant? Did any issues arise that were not anticipated by other researchers who have investigated similar cases? What might be done to bridge the social gap in renewable energy? You will write about these issues in a final reflective paper.

Final Work Product: (1) Final Concept Map + (2) Take-Home Essay + (3) Copy of Part 2 + 3

The final work product for this assignment is an updated version of the concept map and a 1,000-word essay.

Final Concept Map

Revisit the concept map you prepared for class #2. Update any of the boxes to better represent and explain the actors and the issues they've raised. Print it out and turn it in with your essay.

Take-Home Essay

The essay should focus on explanations for the controversy and propose solutions that could resolve either this controversy or future wind power developments. Some questions to guide your essay include the following: what could have been done to bridge the social gap in renewable energy at the Ocotillo Express wind farm site? How are conflicting environmental values present in this controversy? How did the scale of the environmental issue (global, national, local) affect the perceptions of this wind farm? Was the process fair? Was public participation useful? What would you do to resolve the challenges? How do you feel about the project and process? Use the research papers you read in preparation for class #1 and the notes we collectively developed in class. Look for examples from the research. 1,000 words, 1.5 spaced, printed copy in class.

Copy of Part 2 + 3

This is the work completed earlier. Please turn in the stakeholder view essay and original concept use completed in parts 2 and 3 in preparation for class meeting #2. It is fine to mark up with your notes (◘ Fig. 3.2).

Fig. 3.2 Ocotillo Express Wind Farm under construction. Note the spools of copper wire in the foreground in a temporary staging area

3.10 Science and Technology Studies

The subfield of science and technology studies (STS) explores questions that help us understand energy transitions because of engagements with expert knowledge production and public participation. The kinds of researchers that engage in STS include anthropologists, sociologists, public policy researchers, geographers, and people who see their work in the humanities and the arts, and even some engineers.

The STS term "socio-technical imaginaries" is another approach to examining energy futures (Jasanoff & Kim 2009). The idea is that social norms and values are reflected in or are shaped by scientific knowledge. Any sustainable energy strategy will have to be sensitive to the context and geographies specific to energy transitions. Reflexivity is a term used to describe careful reflection and adaptation in policy design to pay attention to how outcomes of energy transitions are developing (Miller & Richter 2014). It will be critical to monitor processes that drive territorialization as renewable or clean energy is deployed. There are already too many reports of "green grabs" for land, water, and other natural resource access that suggest that environmental inequalities and unequal power asymmetries could be reproduced in a green economy. Attention to power dynamics and benefit-sharing arrangements in networks that shape discussions about governance, and access and control of natural resources. These questions extend from indigenous communities near sites of petroleum extraction, such as in the Arctic, to environmental justice communities near sites of petroleum refining (Tysiachniouk & Petrov 2018). Scholars using political ecology suggest their framework allows one to describe the hydro-social territories to see how cultural, political-economic, natural resource flows, and the biophysical environment interact to shape specific outcomes in debates over water (Boelens et al. 2016). The political ecology and STS views attempts to blend explanation across social and ecological actors in these commodity systems.

Some of the richest work on energy and the environment are from environmental historians. These long views on the history and development of these energy infrastructures are important in understanding how such systems are put together today (Cronon 2009). In *Networks of Power*, Thomas Hughes describes the evolution of several electricity networks, showing how they are as much cultural artifacts as outcomes of technological change. As electricity networks diffused from laboratory to a physical grid, it was shaped by institutions, regulatory and legal regimes, and vice versa. Energy historian Christopher Jones (2014) in *Routes of Power* shows how advances in the accessibility of energy supplies, through the development of infrastructures like canals, pipelines, and transmission and distribution systems for electricity, reshaped the geographies of energy production and consumption, putting ever greater distances between those places endowed with energy resources and the cities that draw upon them. Similar to historians, researchers of the history of technology and science specifically, which refer to the social construction of technology, contend that to understand the development of technologies, researchers need to take into consideration the social context in which they are developed, including new legal regimes or cultural norms of users (Bijker 1997).

In the end, "energy needs the social sciences" (Sovacool 2014b, 529). If energy problems truly are social problems with a technical dimension, we need to understand what tools and frameworks the social sciences bring to understanding energy transitions. This chapter offers a sampling of what social science has to say, there is a lot more excellent humanities and social science out there to learn about (White et al. 2015). Furthermore, in energy conversations, it is often economics and policy that are hoisted up as the only considerations from the social world deemed critical to energy transitions. Hopefully, this chapter shows why energy transitions need to take a broader view on what kinds of knowledge is critical to these undertakings.

References

Adger, W. N., Benjaminsen, T. A., Brown, K., & Svarstad, H. (2001). Advancing a political ecology of global environmental discourses. *Development and Change, 32*, 681–715.

United Nations Advisory Group on Energy and Climate Change. (2010). Energy for a sustainable future. New York, April 28, 2010. https://www.un.org/chinese/millenniumgoals/pdf/AGECCsummaryreport%5B1%5D.pdf.

3

Attari, S. Z., DeKay, M. L., Davidson, C. I., & De Bruin, W. B. (2010). Public perceptions of energy consumption and savings. *Proceedings of the National Academy of Sciences, 107*(37), 16054–16059.

Bell, D., Gray, T., & Haggett, C. (2005). The 'social gap' in wind farm siting decisions: Explanations and policy responses. *Environmental Politics, 14*(4), 460–477.

Bellamy Foster, J. (2008). Marx's *Grundrisse* and the ecological contradictions of capitalism. In M. Musto (Ed.), *Karl Marx's Grundrisse*. London: Routledge.

Bennett, J. (2009). *Vibrant matter: A political ecology of things*. Durham: Duke University Press.

Berger, G., Flynn, A., Hines, F., & Johns, R. (2001). Ecological modernization as a basis for environmental policy: Current environmental discourse and policy and the implications on environmental supply chain management. *European Journal Social Science, 14*, 55–72.

Bidwell, D. (2013). The role of values in public beliefs and attitudes towards commercial wind energy. *Energy Policy, 58*, 189–199.

Bijker, W. E. (1997). Of bicycles, bakelites, and bulbs: Toward a theory of sociotechnical change. MIT Press, Cambridge.

Blaikie, P., & Brookfield, H. (1997). *Land degradation and society*. London: Routledge.

Boelens, R., Hoogesteger, J., Swyngedouw, E., Vos, J., & Wester, P. (2016). Hydrosocial territories: a political ecology perspective.

Bryant, R. L., & Goodman, M. K. (2004). Consuming narratives: The political ecology of 'alternative' consumption. *Transactions of the Institute of British Geographers, 29*(3), 344–366.

Bullard, R. D. (1996). Environmental justice: It's more than waste facility siting. *Social Science Quarterly, 77*(3), 493–499.

Cole, L., & Foster, S. (2001). *From the ground up: Environmental racism and the rise of the environmental justice movement*. New York: New York University Press.

Caniato, F., Caridi, M., Crippa, L., & Moretto, A. (2012). Environmental sustainability in fashion supply chains: An exploratory case based research. *International Journal of Production Economics, 135*(2), 659–670.

Carroll, J., Lyons, S., & Denny, E. (2014). Reducing household electricity demand through smart metering: The role of improved information about energy saving. *Energy Economics, 45*, 234–243.

Cook, I., & Crang, P. (1996). The world on a plate culinary culture, displacement and geographical knowledges. *Journal of Material Culture, 1*(2), 131–153.

Clark, H., (2011). Sustainable energy access critical for development in Africa. United Nations Development Programme. https://www.undp.org/content/undp/en/home/ourperspective/ourperspectivearticles/2011/12/29/sustainable-energy-access-critical-for-development-in-africa-helen-clark.html.

Cronon, W. (2009). *Nature's metropolis: Chicago and the great west*. New York: WW Norton & Company.

Cutter, S. L., Barnes, L., Berry, M., Burton, C., Evans, E., Tate, E., & Webb, J. (2008). A place-based model for understanding community resilience to natural disasters. *Global Environmental Change, 18*(4), 598–606.

Diamond, S. (1998). *Not by politics alone: The enduring influence of the Christian Right*. New York: The Guilford Press.

Dietz, T., Gardner, G. T., Gilligan, J., Stern, P. C., & Vandenbergh, M. P. (2009). Household actions can provide a behavioral wedge to rapidly reduce US carbon emissions. *Proceedings of the National Academy of Sciences, 106*(44), 18452–18456.

Dietz, T., Stern, P. C., & Weber, E. U. (2013). Reducing carbon-based energy consumption through changes in household behavior. *Daedalus, 142*(1), 78–89.

Dryzek, J. (1997). *The politics of the earth*. Oxford: Oxford University Press.

European Commission. (2011). Addressing the rebound effect. Report to the European Commission. April 26, 2011. https://ec.europa.eu/environment/eussd/pdf/rebound_effect_report.pdf.

Gereffi, G., Humphrey, J., & Sturgeon, T. (2005). The governance of global value chains. *Review of International Political Economy, 12*(1), 78–104.

Gereffi, G., Humphrey, J., & Kaplinsky, R. (2001). Introduction: Globalisation, value chains and development. *IDS Bulletin, 32*(3), 1–8.

Gillingham, K., & Palmer, K. (2014). Bridging the energy efficiency gap: Policy insights from economic theory and empirical evidence. *Review of Environmental Economics and Policy, 8*(1), 18–38.

Graziano, M., & Gillingham, K. (2014). Spatial patterns of solar photovoltaic system adoption: The influence of neighbors and the built environment. *Journal of Economic Geography, 15*(4), 815–839.

Habermas, J. (1981). New social movements. *Telos, 49*, 33–37.

Hess, D. J. (2018). Energy democracy and social movements – A multi-coalition perspective on the politics of sustainability transitions. *Energy Research & Social Science, 40*, 177–189.

Holdren, J. P., Smith, K. R., Kjellstrom, T., Streets, D., Wang, X., & Fischer, S. (2000). *Energy, the environment and health*. New York: United Nations Development Programme.

Hopkins, T. K., & Wallerstein, I. (1986). Commodity chains in the world-economy prior to 1800. *Review (Fernand Braudel Center)*, 157–170.

IEA (International Energy Agency). (2016). World energy outlook. https://www.iea.org/newsroom/news/2016/november/world-energy-outlook-2016.html.

Jasanoff, S., & Kim, S. H. (2009). Containing the atom: Sociotechnical imaginaries and nuclear power in the United States and South Korea. *Minerva, 47*(2), 119.

Jevons, W. S. (1906). *The coal question: An inquiry concerning the progress of the nation, and the probable exhaustion of our coal-mines*. London: Macmillan.

Jones, V. (2007). Is the environmental movement too white? April 17, 2007. http://www.commondreams.org. Accessed 3 June 2007.

Jones, C. F. (2014). *Routes of power*. Cambridge: Harvard University Press.

Kitschelt, H. P. (1986). Political opportunity structures and political protest: Anti-nuclear movements in four democracies. *British Journal of Political Science, 16*(1), 57–85.

Liechti, F., Guélat, J., & Komenda-Zehnder, S. (2013). Modelling the spatial concentrations of bird migration to assess conflicts with wind turbines. *Biological Conservation, 162*, 24–32.

Litfin, K. (1994). *Ozone discourses: Science and politics in global environmental cooperation*. New York: Columbia University Press.

Lockie, S., & Kitto, S. (2000). Beyond the farm gate: Production-consumption networks and Agri-food research. *Sociologia Ruralis, 40*(1), 3–19.

Luke, T. W. (2005). The death of environmentalism or the advent of public ecology? *Organization & Environment, 18*(4), 489–494.

MacKay, D. (2009). *Sustainable energy without the hot air*. Cambridge: Cambridge University Press. http://www.withouthotair.com/.

Maantay, J. (2007). Asthma and air pollution in the Bronx: Methodological and data considerations in using GIS for environmental justice and health research. *Health & Place, 13*(1), 32–56.

Bernardino, J., Bevanger, K., Barrientos, R., Dwyer, J. F., Marques, A. T., Martins, R. C., & Moreira, F. (2018). Bird collisions with power lines: State of the art and priority areas for research. *Biological Conservation, 222*, 1–13.

Mather, C. (1999). Agro-commodity chains, market power and territory: Re-regulating South African citrus exports in the 1990s. *Geoforum, 30*(1), 61–70.

McAdam, D. (1982). *Political process and the development of black insurgency*. Chicago: University of Chicago Press.

Miller, B. (2000). *Geography and social movements: Comparing antinuclear activism in the Boston area*. Minneapolis: University of Minnesota Press.

Miller, C. A., & Richter, J. (2014). Social planning for energy transitions. Current Sustainable/Renewable Energy Reports, 1(3), 77–84.

Mulvaney, D. (2019). *Solar power: Innovation, sustainability, and environmental justice*. Oakland: University of California Press.

O'Brien, K. L., & Leichenko, R. M. (2000). Double exposure: Assessing the impacts of climate change within the context of economic globalization. *Global Environmental Change, 10*(3), 221–232.

O'Connor, M. (1994). The second contradiction of capitalism. *The Material/Communal Conditions of Life, 5*(4), 105–114.

Pasqualetti, M. J. (2001). Wind energy landscapes: Society and technology in the California Desert. *Society & Natural Resources, 14*(8), 689–699.

Pasqualetti, M. J. (2011). Opposing wind energy landscapes: A search for common cause. *Annals of the Association of American Geographers, 101*(4), 907–917.

Pellow, D., & Park, L. (2002). *Silicon Valley of dreams: Environmental injustice, immigrant workers, and the high-tech global economy*. New York: New York University Press.

Phadke, R. (2011). Resisting and reconciling big wind: Middle landscape politics in the New American West. *Antipode, 43*(3), 754–776.

3

Piven, F., & Cloward, R. (1979). *Poor peoples movements: Why they succeed and fail*. New York: Random House.

Robbins, P. (2004). *Political ecology: A critical introduction*. London: Blackwell.

Robbins, P. (2011). *Political ecology: A critical introduction*. New York: John Wiley & Sons.

Robić, S., & Ančić, B. (2018). Exploring health impacts of living in energy poverty: Case study Sisak-Moslavina County, Croatia. *Energy and Buildings, 169*, 379–387.

Schurman, R. (2004). Fighting "Frankenfoods": Industry opportunity structures and the efficacy of the anti-biotech movement in Western Europe. *Social Problems, 51*(2), 243–268.

Sen, A. (1999). *Development as freedom*. Oxford: Oxford University Press.

Shrader-Frechette, K. (2002). *Environmental justice: Creating equality, reclaiming democracy*. Oxford: Oxford University Press.

Smith, N. (2010). *Uneven development: Nature, capital, and the production of space*. Athens: University of Georgia Press.

Sovacool, B. K. (2014a). Cornucopia or curse? Reviewing the costs and benefits of shale gas hydraulic fracturing (fracking). *Renewable and Sustainable Energy Reviews, 37*, 249–264.

Sovacool, B. K. (2014b). Energy studies need social science. *Nature, 511*(7511), 529.

Sovacool, B. K., Ali, S. H., Bazilian, M., Radley, B., Nemery, B., Okatz, J., & Mulvaney, D. (2020a). Sustainable minerals and metals for a low-carbon future. *Science, 367*(6473), 30–33.

Sovacool, B. K., Hook, A., Martiskainen, M., Brock, A., & Turnheim, B. (2020b). The decarbonisation divide: Contextualizing landscapes of low-carbon exploitation and toxicity in Africa. *Global Environmental Change, 60*, 102028.

Spaargaren, G., & Mol, A. P. (1992). Sociology, environment, and modernity: Ecological modernization as a theory of social change. *Society & Natural Resources, 5*(4), 323–344.

Stern, P. C., Sovacool, B. K., & Dietz, T. (2016). Towards a science of climate and energy choices. *Nature Climate Change, 6*(6), 547.

Stock, R., & Birkenholtz, T. (2019). The sun and the scythe: Energy dispossessions and the agrarian question of labor in solar parks. *The Journal of Peasant Studies*, 1–24.

Swyngedouw, Erik. (1997). "Excluding the other: The production of scale and scaled politics." in Geographies of economies, R. Lee, and J. Wills (eds.), London: Arnold Press.

Szasz, A. (1994). *Ecopopulism: Toxic waste and the movement for environmental justice*. Minneapolis: University of Minnesota Press.

Tabassum-Abbasi, Premalatha, M., Abbasi, T., & Abbasi, S. (2014). Wind energy: Increasing deployment, rising environmental concerns. *Renewable and Sustainable Energy Reviews, 31*, 270–288.

Tysiachniouk, M. S., & Petrov, A. N. (2018). Benefit sharing in the Arctic energy sector: Perspectives on corporate policies and practices in Northern Russia and Alaska. *Energy Research & Social Science, 39*, 29–34.

Van der Horst, D. (2007). NIMBY or not? Exploring the relevance of location and the politics of voiced opinions in renewable energy siting controversies. *Energy Policy, 35*(5), 2705–2714. https://doi.org/10.1016/j.enpol.2006.12.012.

Walker, G., & Cass, N. (2007). Carbon reduction, "the public" and renewable energy: Engaging with socio-technical configurations. *Area, 39*(4), 458–469.

Watts, M. (2000). Political ecology. *A companion to economic geography, 257*, 274.

White, D., Rudy, A., & Gareau, B. (2015). *Environments, natures and social theory: Towards a critical hybridity*. London: Macmillan International Higher Education.

World Health Organization. (2009). Exposure to air pollution: A major public health concern. https://www.who.int/ipcs/features/air_pollution.pdf.

Energy and the Environment I: Fossil Fuels

Contents

D. Mulvaney, *Sustainable Energy Transitions*,
https://doi.org/10.1007/978-3-030-48912-0_4

4

Learning Goals

At the end of this chapter, readers should be able to:

- understand the nexus of energy challenges and relevant economic, social, and environmental issues;
- describe the spatial distribution of conventional energy sources;
- analyze the relative energy use in US to other nations, and the forces that shift the mix of energy sources over time;
- describe basic principles to improve efficiency and design of energy delivery, recognize opportunities to reduce energy consumption, and promote sustainability;
- assess basic economic, government policy, and social equity dimensions of energy options; and
- utilize tools to evaluate an energy option and assess alternatives.

Overview

The goal of this chapter on energy and the environment is to review the basic impacts of human civilization's quest for natural resources and energy and how these developments impact socio-ecological systems. Each of the major fossil fuel sources—natural gas, petroleum, coal, tar sands, and oil shale—is described in terms of major impacts, current and historical commodity chains, and impacts to communities and ecosystems. Recent advances in unconventional shale oil and gas production, and even less conventional and emerging energy extraction techniques, such as tar sands, oil shale (kerogen oil), and methane hydrates, are detailed through problem sets and writing assignments on existing energy commodities and infrastructure. These examples draw on lessons from prior chapters regarding unit transformations and calculating energy/power flows. Readers will learn to estimate potential production of various energy sources from Earth's resources and describe the key pollution impacts from particular power plants of given sizes, capacity, and emissions factors, and interpret the impacts on the environment as well as suggest and design mitigations and ways to minimize and avoid these impacts.

4.1 Energy and Society

Transforming energy into the services that power everyday life is a remarkable achievement of human civilization. The harnessing of energy, for heat, light, and motion, has led to better lives for most of human civilization, particularly since the industrial revolution. At the same time, energy extraction, transport, production, and use shape and impact society and the environment in different and uneven ways.

Energy production and use reshape society and the environment in significant ways that are not easy to reduce. Per capita energy use has increased orders of magnitude over the years since the dawn of sedentary civilization, but so have vital statistics for human quality of life. Childhood mortality, female life expectancy, and the number of people who experience malnutrition have all improved as energy use per capita has increased. At the same time, the environmental impacts of energy systems have also grown significantly. Impacts to air, water, land, and even people occur from decisions made about where and how much energy is extracted or produced in various places. This chapter dives deeper into environmental impacts of energy, accepting as a starting position that the increase in energy use in aggregate has improved social outcomes overall, but that there are critical challenges.

Energy access, first and foremost, is the interaction with energy that matters most if some place wants a population that is well-nourished and with high ages for female life expectancy. Extractive industries can complicate claims about the general improvement of the human condition with energy access. As energy extraction increased, so did the mark of human activity on the landscape. Historian Lewis Mumford in his 1934 treatise *Technics and Civilization* noted that "carboniferous capitalism" relies on the "accumulated wealth" of carbon-based energy that took millennia to amass (Mumford 1934). The fossil fuels that power civilization today are the products of long ecological and geological processes—hundreds of millions of years—and its use slowly diminishes a stock of finite low-entropy resources. Mumford distinguished carboniferous capitalism from other eras where human civilization acquired flows of energy from the sun, mainly from wood and agricultural biomass. Mumford argued that carboniferous capitalism's "habituation to wreckage and debris" with "disregard for a balanced mode of production and consumption" would strip human civilization of the energy resources upon which it is built. The upshot of Mumford's concern can be connected to Georgescu-Roegen's ideas about entropy and the economy, described earlier, where each transformation of resources degrades the quality of energy and drains stocks of natural resources.

We can think of useful energy in the forms of stocks and flows. Stocks of energy and natural resources can be drawn down, while flows of energy and natural resources are constantly replenishing.

The addition of carbon into the atmosphere by way of combustion of fossil fuels is also a major intervention in the carbon cycle. Scientific research is teaching us that adding carbon to the atmosphere is changing the planet's climate. The greenhouse effect is the result of accumulated levels of gases like carbon dioxide, methane, and nitrous oxide that trap solar energy in the atmosphere. Solar energy delivered to Earth is absorbed and re-radiated out as heat in the form of long-wave radiation. The effect was first widely publicized by the Swedish scientist Arrhenius in the late 1800s. But it was not until scientific evidence from various sources, including tree rings to ice cores by the 1960s, that the effect has become widely accepted. Switching away from carbon-based energy stocks (coal, petroleum, natural gas, even uranium some argue) to renewable flows of energy is now seen as one of humanity's most pressing challenges.

Ecological economists argue that there is a thermodynamic basis for energy transitions toward renewable flows of energy. Stocks of energy resources are finite and eventually depleted. But constantly replenishing energy resources like the sun, wind, and geothermal energy continue so long as those Earth system processes continue. Most natural resources have environmental constraints and limits to availability. Geologic limits describe the amount of certain metals in the Earth's crusts. Considering economic constraints, some geologically present resources are not economic to extract. The twenty-first century is an era of declining ore grades, meaning that there are fewer and fewer easily and readily extractable natural resources over time. Copper ores, for example, may have been on the order of 3–4% several decades ago. Now, copper is mined at concentrations less than 1%. The cheapest and most highly concentrated wastes are usually found first. This is prompting discussions about deep-sea and space mining for critical or scarce metals, and the need to protect them from exploitation (Christensen et al. 2019).

While some of the limits are due to scarcity of cheap or cost-effective natural resources. Some of these limits are related to the capacity of environmental systems to absorb pollution. There is a limit according to scientists to how much greenhouse gas pollution can enter the atmosphere without climate change feedbacks. Likewise, there are sometimes water resource limits that effect energy and natural resource extraction, or limits on the capacity of an environment to absorb waste.

Air pollution is a widely known impact of energy use from tailpipe emissions from automobiles to indoor air pollution from indoor flames. A total of 90% of air pollution—nitrogen oxides (NO_x), sulfur oxides (SO_x), particulate matter (PM), heavy metals like lead, and volatile organic compounds (VOCs)—is caused by combustion according to the US Environmental Protection Agency (EPA), and the effects are usually borne by poor and minority communities, often the most vulnerable to begin with. The Great Smog of London in 1952 killed hundreds of people, and a deadly air pollution inversion in Donora, Pennsylvania, helped usher in new rules for air pollution and air quality standards. Since the air pollution laws were passed in the 1970s (and earlier in some US states), there have been significant improvements in air quality. Globally, the adoption of air pollution law is more uneven. As other countries develop and industrialize, they appear to pass through a phase of development with significant air pollution. Megacities like Beijing, China, Sao Paola, Brazil, and Delhi, India, have notoriously poor air quality. But even advanced regional economies, like California, are challenged by air quality problems.

Ozone in the upper reaches of the atmosphere blocks otherwise harmful ultraviolet light. But this is far above the reaches of the air that humans breathe. When ozone is produced in the air at the ground level, it is a lung irritant and can cause significant heart and lung damage. Ozone is not a primary pollutant. It is produced from nitrogen oxides (NO_x) via a photochemical reaction with the sun. The chemistry of the breakdown of nitrogen oxides to ozone is as follows.

Definition

$N_2 + O_2$ + high temperatures $\rightarrow$ 2 NO	Nitric oxide, primary pollutant
$2\ NO + O_2 \rightarrow 2\ NO_2$	Nitrogen dioxide
NO_2 + higher energy sunlight $\rightarrow$ NO + O	NO_x + oxygen radical
$O + O_2 \rightarrow O_3$	Ground-level ozone

Some communities where there is a lack of monitoring, no trust in data, or no regulation have taken it upon themselves to conduct citizen science projects to collect air pollution data near refineries or fracking sites to understand the patterns and produce their own observations (Gabrys 2017). Other examples include "bucket brigades" where there are water pollution issues that impair waterways and drinking water.

4.2 Coal

Coal is an important resource in human history. The primary energy resource provided more electricity generation than any other source since the industrial revolution. For industry, coal made for better metallurgy, and coal is used to produce steel and for making semiconductors still today. Throughout the history of mining coal for electricity and steel production, mining has been dangerous work. The resource is the subject of stories today around carbon pollution, but in the past, the emphasis was on labor standards and occupational safety of miners. Coal deserves acknowledgment for the role it plays in bringing billions of humans out of poverty. The benefits of electrification and associated employment in communities that depend on the resource for economic activity. Offering retraining and other job-related skills are part of energy transition conversations that seek to offer some opportunity to communities and workers that may be impacted by climate policies. Some retraining might allow them to move into adjacent fields. One study of energy transition in coal found that solar power is more suitable than wind to replace lost coal jobs because of the resources available in coal regions and communities (Pai et al. 2020).

The coal resource mainly is a product of processes that occurred during the carboniferous period—360 to 286 million years ago, when the Earth's climate was warmer and more humid. Vegetation and other sources of carbon are sedimented upon, starved of oxygen, and decay under pressure, at high heat, and over time, to form coal. Coal is produced from the cellulose and lignin in plants. When these materials are buried in sediments, they decay to a class of materials called vitrinites. The youngest form of coal is peat, a material produced from the compression of sedimentary layers of rock, soil, and sand. Many parts of the world still use peat as a heating fuel, though it has moisture and other impurities. Eventually, under more heat, pressure, and the patience of a few million years, peat forms lignite, often called "brown coal." Most brown coal reserves come from the Quaternary period—starting over two million years ago.

Further compression and time form subbituminous and bituminous coal. Where pressures are greatest, a very carbon-rich form of coal called anthracite is formed. Anthracite is an extremely uncommon coal type that forms where there are significant plate collisions and carbon-rich ecosystems. The anthracite mines of eastern Pennsylvania, US, for example, were formed when the plate that carries present-day Africa collided with the east coast of North America. The collision occurred a long time after a sedimentary basin with carbon-rich ecosystem was deposited.

As coal ages and is pressurized and heated, the fixed carbon content of coal increases. Peat has the lowest carbon content, followed by lignite, subbituminous,

Fig. 4.1 A coal seam from the Kaiparowits Plateau in southern Utah

bituminous, and anthracite. The amount of volatile matter and moisture follows the opposite trend with very little of those impurities in anthracite. Anthracite with the highest carbon content is called meta-anthracite, while semi-anthracite is a bit lower in fixed carbon content. These grades also predict the energy content or calorific value of coal as bituminous coal and anthracite have the highest energy content. As the carbon content increases, the heating value or energy density is reduced slightly. So high volatile bituminous coal and semi-anthracite have similar energy densities (Fig. 4.1).

Compared to the renewables discussed later in this chapter, coal has a very high energy density. Lignite resources can have densities that compare to dried dung or air-dried wood (10–20 megajoules per kilogram (MJ/kg)); bituminous coal can range from 20 to 26 (MJ/kg), whereas anthracite range from 27 to 30 MJ/kg (Smil 1991).

Definition

Energy Density—a metric that represents a quantity of energy per unit area or unit volume.

Centuries ago, coal production and use were primarily for metallurgy (making metal) and heating homes. By the time it was used to power electricity and power prime movers, the scale of power output increased significantly. Thomas Newcomen's steam engine from 1730, used to pump water out of coal and iron mines, were on the order of 10 kilowatts (kW) and smaller. They were also very inefficient. Newcomen's steam engine had an efficiency of 0.5%. James Watt's largest steam engine (built in the 1800s) was around 100 kW. Modern gas- and steam turbines are many orders of magnitude higher, from 100,000 kW to 1,500,000 kW.

Geographically, coal production is concentrated in a few places, but it is spread throughout the world wherever rocks of viable age and content are present. A total of 7.3 billion tons of coal are mined globally every year. China (2300 million tons (Mt)), India (708 Mt), the US (672 Mt), Australia (503 Mt), and Indonesia (461 Mt) are leading coal-producing countries, with also very high reserves still in the ground (World Energy Council 2017). Regions and countries without much coal include Africa, Japan, and the West Coast of North America.

In the US, there are several important coal-bearing regions, including the (1) Intermountain West, which includes the Rocky Mountain and Colorado Plateau regions, and the states Wyoming, Montana, Colorado, Utah, New Mexico, and Arizona; (2) Appalachia, including the states of West Virginia, Kentucky, Pennsylvania, Tennessee, and Mississippi; and the (3) Interior region, where coal is scattered from the Gulf coast to the Illinois Basin states, including Texas, Louisiana, Illinois, Michigan, Indiana, and other states to a lesser extent. The Powder River Basin surface mines in Wyoming supply up to 40% of US coal production annually, mainly from 13 large strip mines near Gillette. With 40 billion tons of extractable coal remaining, the Powder River Basin will be the major supplier of coal to the US for decades to come. Appalachia, with the high mine-mouth costs of mountaintop removal, and the high sulfur content in their coal, has seen a dramatic decline over the past four decades. The two broadest categories of coal are thermal coal and metallurgical coal, the latter of which is used to make silicon metal and steel (◘ Fig. 4.2).

Coal usually requires transport from where it is mined to its site of use. The top coal power electricity-generating states in the US are Texas, Indiana, Ohio, Illinois, and Pennsylvania. Interestingly, few of these states get much of their coal needs from their own states, as many of these plants are fed with coal from the American Interior West. Over 90% of coal is used in the electricity sector, but coal is also used in other industrial sectors, including industrial processes to make heat and steam, for coking metallurgical silicon and steel, and to a lesser extent for home and commercial heating applications. Sometimes, this coal is distinguished as metallurgical coal. The US exports a large portion of this kind of coal to other countries.

In China, much of the coal production comes from regions in the north of the country. Most of these coal-fired power plants are located in the coastal regions; the development of coal as a cheap energy source required the development of railway infrastructure to move coal at low cost. Much as the rest of the world, China's coal industry is undergoing rapid mechanization and using few workers per unit energy mined. By the time coal ends its reign as a major source of energy, China will have consumed about 50% of global coal ever produced (Wang and Li 2016).

Extracting coal from the Earth is done by subterranean methods or means that terraform landscapes by removing large volumes of Earth. Throughout much of

Fig. 4.2 Open-pit strip mining in the Powder River Basin, Wyoming, from space. (Image courtesy NASA)

the early years of the industry, most coal extraction occurred by hand by workers in underground mines. This can be very dangerous work. In these early days of coal mining, there were very high rates of worker injury and death in this industry as coal miners were exposed to (1) "blackdamp"—elevated levels of CO_2 (which can cause suffocation if it displaces fresh air), (2) "afterdamp"—CO (a blood poison when exposure levels are too high), and (3) "firedamp"—an explosive mix of methane (CH_4), coal dust, and hydrogen sulfide (H_2S), any of which can also result in mine fires and explosions. These underground dangers coupled with the potential for mine and mine shaft collapses made mining one of the most dangerous professions throughout much of the industrial revolution.

Firedamp caused three of the world's most fatal coal mine disasters: the Darr Mine disaster in Pennsylvania (US) and Gresford and Abercarn colliery disasters in Wales (UK), which led to the deaths of 239, 266, and 268 miners, respectively, when lamps used by the miners ignited the deadly mixture of gases. Nearly 4,000 workers died annually in the coal mines of China in the 2000s, but more recently, deaths in the coal mining industry have fallen to record lows (CLB 2019).

In the US, there were nearly 800,000 workers working in coal mines in 1920, falling to 200,000 by 1960, to less than 50,000 in 2015. Mechanization has resulted in fewer but larger mines. Mountaintop removal (MTR) is a coal mining practice using large equipment to remove topsoil, vegetation, and rock—so-called overburden—to expose large seams of coal. Draglines are massive cranes that take the material left over from MTR, the rocks, tree debris, and other overburden. The trucks used to move coal and overburden have wheels over ten feet in diameter, some a lot larger. Mechanized production processes are better for workers than the small underground mines. Among the more concerning impacts of MTR are severe water quality and riparian impacts. MTR begins with benches cut into the hillsides to allow access to places to set off explosives. In the Appalachia region of Eastern North America, MTR mainly occurs in mixed mesophytic forests, important seed beds for the continental forests. Once the coal seams are revealed, they are removed to the next stage of processing. Most overburden is moved by dragline into low-lying areas or used to reconstruct contours. The US Surface Mining Control and Reclamation Act of 1977 requires coal mines on federal land be returned to the "approximate to the original contour" (Udall 1978). Many states in the US require that companies make some concerted effort to return the site to its prior condition—often referred to as reclamation. Reclamation remains challenging as, often, the soils and seedbeds have been removed.

In total, 60% of particulate matter (PM) is from coal, 45% of SO_2, 30% of NO_x, and 80% of mercury (EPA 2011). Lesser amounts of cadmium, antimony, dioxins, furans, and lead emissions can also occur from coal-fired generators. When sulfur dioxide and other sulfur compound (SO_x) emissions from coal are mixed with water in the hydrosphere, acid rain is formed. Weather systems can deposit acid rain far away and harm aquatic life, soils, and vegetation sensitive to pH.

Mercury pollution hot spots are usually found near either mining activities or coal-fired power plants. Mercury is a contaminant found in coal, and after combustion, it is deposited in the environment as methyl-mercury emitted from coal-fired generation. These heavy metals biomagnify up the food chain, moving from krill, to krill-eating fish, through piscivorous fish, and ultimately the human food supply.

Other air pollutants from burning coal besides NO_x and PM include dioxins, furans, arsenic, selenium, cadmium, and molybdenum. Pollution control equipment can significantly reduce these emissions. The best available technologies used in air pollution control for fossil fuel combustion include electrostatic precipitators, filters, scrubbers, and fluidized bed absorbers.

Fly ash is a solid waste product from burning coal that contains highly toxic heavy metals and furans. Fly ash made into a slurry is moved into ponds located on site with the power plant and is often held back from other water bodies by earthen dams. These ponds are linked to dredge cells where the materials are evaporated and solids settle to the bottom before being sent off for permanent disposal at a waste treatment facility or landfill. These dams have been known to break open and spill liquid fly ash into water bodies and neighborhoods. One well-known case came near the Kingston coal-fired generator in the Tennessee Valley Authority's (TVA) service area, when millions of gallons of toxic sludge poured into a neighborhood and the Emory River, a tributary of the Chattanooga River (New York Times 2009). Some other uses for fly ash include bricks and wallboard, so long as the metals can be sealed or removed.

Some environmental impacts are related to inputs to the coal supply chain. A coal-cleaning chemical 4-methylcyclohexane methanol spilled into West Virginia's Elk River, which led to over 100,000 people losing drinking water in the state for many weeks.

Coalbed methane is another energy source developed from coal fields, but that is actually natural gas, mostly methane. The three major coalbed methane fields in the US are in the Powder River, San Juan, and Black Warrior Basins. Coalbed methane is also widely practiced and controversial in Australia.

Carbon capture and storage (CCS) technologies will be critical to any future uses of fossil fuels. There are numerous ways to capture and store carbon. Many CCS technologies require converting coal into syngas—short form for synthetic gas—that comprises hydrogen, carbon monoxide, and carbon dioxide. Syngas is one of the first uses of coal as an energy resource, as the fuel was produced and distributed to light city gas lamps, and finally delivered to home for mainly lighting.

The reaction to produce syngas is:

$$\text{Coal}\left(\text{mostly carbon, hydrogen}\right) + O_2 + H_2O \rightarrow H_2 + CO$$

Integrated gasification combined cycle (IGCC) plants proposed for CCS also produce syngas. "Clean coal" technologies optimistically would remove carbon from combustion emissions. Carbon captured would be delivered to porous rocks, saline aquifers, or depleted oil reservoirs, where CO_2 is used for enhanced oil recovery (EOR). Combined cycle (CC) technologies take waste heat from combustion to power a second boiler system and turbine.

Coal can also be converted into diesel fuel. This process is known as coal liquefaction or coal-to-liquids. First, syngas is produced. Second, a technique called the Fischer-Tropsch process converts syngas in the presence of a catalyst into long chains of diesel molecule, with water as a by-product. Many countries from the US to China have explored coal-to-liquids to help alleviate the high costs associated with importing oil and natural gas.

The Fischer-Tropsch process is:

$$(2n+1)H_2 + n\,CO \rightarrow C_nH_{(2n+2)} + n\,H_2O$$

CCS faces several key hurdles. First, there are unresolved issues in the modeling, monitoring, and verification of CCS options and strategies to ensure they are effective. Social arrangements such as property rights and liability will need attention as well. Several states have created new property rights in pore space, the vacant openings where oil or gas has been extracted, and where CO_2 may be stored. Another question concerns who will be liable if leakage occurs or, worse, if people are killed from the release of large amounts of CO_2 that displace air. There are unresolved issues with site licensing and monitoring. Does the stored carbon stay put? Are compensation arrangements between companies burning CO_2 and communities needed? There will be long-term costs to the public because if companies go bankrupt, no one can be held responsible.

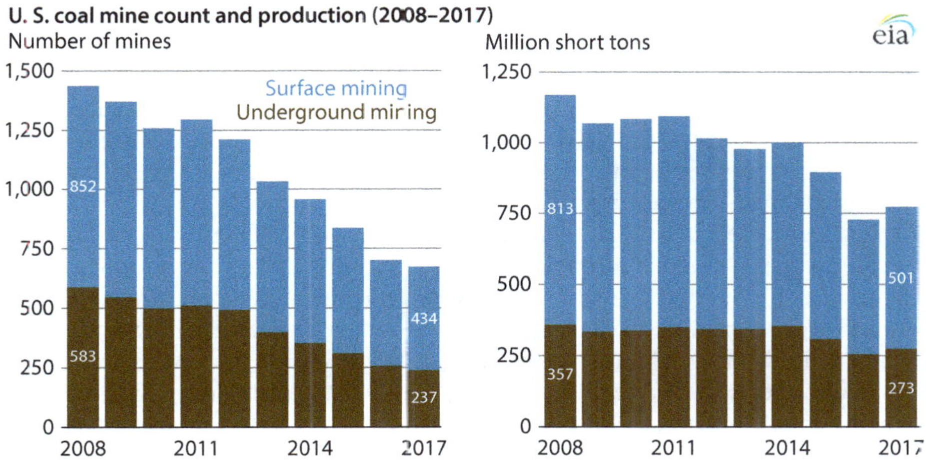

Fig. 4.3 Half of US coal mines operating in 2008 were closed by 2017

One form of CCS widely investigated is Bioenergy with Carbon Capture and Storage (BECCS). BECCS is receiving increased attention these days as a potential low-carbon energy option. BECCS involves growing highly productive energy crops that take up carbon from the atmosphere, producing electricity or liquid fuels like ethanol, and storing any carbon captured in the process underground.

Another CCS option is reforestation or afforestation, where again carbon is taken from the atmosphere, but this time sequestered by trees and associated biomass. The capture and storage of carbon would most likely require some incentive like a carbon tax to ensure that there is a market for the carbon (Fig. 4.3).

Softening the impacts of energy transitions in the coal industry are widely discussed. For example, there have been suggestions for grant programs to support retraining in clean and renewable energy industries for workers. In Colorado, a policy was put into place to fund the training of workers for new industries as a result of actions to encourage the early closure of coal-fired power plants (The Energy Transition Show 2019).

4.3 Natural Gas

Natural gas uses the adjective "natural" to distinguish it from other gases like coal-gas, syngas, or "town" gas; some say it should be called "fossil gas." As a final fuel, natural gas is mainly methane (CH_4) and often referred to as "dry" natural gas. "Wet" natural gas is the collection of hydrocarbons extracted from the Earth. That material includes water, ethane (5–15%), propane (<5%), and butane (<5%), in addition to methane (70–90%), with CO_2, N_2, and H_2S also found as impurities. How clean the gas is and how much of these higher carbon constituents are found can greatly affect how much value any particular extractive site can be.

In 1970, crude oil supplied 46% of the world's energy, 43% in the US, and 50% in the European Union (EU), while natural gas only supplied 18% of the world's energy (Smil 2012). Coal has always been the dominant energy source for electricity. Yet, today, due to advancements in hydraulic fracturing, natural gas is the single-most important fossil fuel resource, supplying up to 24% of the world's energy in 2010 (Smil 2012). From 2013 to 2017, shale gas production in the US rose from 8000 to over 17,000 billion cubic feet per year. The potential reserves are over 800 trillion cubic feet—enough to last for decades (Mooney 2011). Though by 2020, the industry was experiencing a debt crisis.

As a resource, natural gas is found directly, or with oil. Natural gas is derived from kerogens—organic compounds in sedimentary rocks. The proportions of natural gas to oil at a particular site depend on originating kerogen mix, with oil derived from a material called sapropel, and natural gas derived from vitrinites. Methane is also formed biologically from methogens (biogenic methane). Natural gas and oil fields are scattered throughout the world's sedimentary rocks, and the fields are often referred to as "plays."

Natural gas is described as either "wet" or "dry" natural gas. Wet natural gas contains all of the hydrocarbons that are extracted, including ethane, butane, propane, and hexane. Dry natural gas is the finished methane product that is delivered to homes and businesses. These are very important distinctions as sometimes metrics confound the two. Try to determine whether the research or analysis means natural gas produced at a well or whether they mean molecule methane when natural gas is invoked. Hopefully, analysts are following this distinction.

Dry natural gas—methane—is a versatile fuel used to make electricity, heat buildings and homes, provide warm water, and even power some kinds of vehicles and buses. Natural gas is also used in several industrial processes. The Claude-Haber process to make fertilizer has been used since 1910. This amounts to 5% of natural gas use (energy + feedstock for H_2)—1–2% of total global energy use annually. Most natural gas is used for heat or to boil water to generate electricity. Polyethylene is made from wet natural gas (ethane) by companies like Exxon, Royal Dutch Shell, and Dow. Ethane for polyethylene is one of the most profitable products made from wet natural gas. The importance of ethane to the polyethylene and polyolefins industries make it likely that a low-carbon energy transition will likely coincide with a green chemistry revolution in bioplastics.

A few basic energy principles related to natural gas

- 1 Btu will raise the temperature of 1 lb of water 1 °F
- 1 cubic foot of gas will raise 1,000 lbs of water 1 °F

Question: How many Btu are in 1 cubic foot of gas?

Answer: 1,000 btu of natural gas boils about 1 gallon of water.

A few other considerations related to energy and carbon metrics for natural gas.

0.12 lb CO_2/ft^3

116.4 lb CO_2/MMbtu

Typically, natural gas and propane are sold in therms or cubic feet.

1 MMbtu = 1,000 Mbtu = 1 million btu = 10 therms

Most conventional gas production is by drilling a well and drawing gas and liquids from the depths of a natural gas field. Sometimes natural gas is associated with oil in proportions that depend on the particular reservoir under development and its underlying geology. These can be either onshore or offshore wells. Unconventional means of natural gas production include tight sand gas or tight gas, gas-rich shale or shale gas, gas hydrates, and coalbed methane. The geological formations where natural gas is extracted can be near Earth's surface, to several miles deep. The "plays" can be layered atop each other like a cake. For example, near Pennsylvania, Ohio, and West Virginia in the US, the Marcellus, Utica, and Devonian shale layers that contain natural gas occur all at different depths. Similarly, along the Texas-Louisiana region, the Travis Peak, Bossier, Cotton Valley, and Gilmer line are tight gas plays that are found at different depths (Fig. 4.4).

In the mid-2000s, as hydraulic fracturing ("fracking') began to take root in an unprecedented rollout of a new extraction technique, many countries like the US were planning for a future of natural gas imports by building liquefied natural gas (LNG) import terminals. In 2005, there were dozens of LNG import terminals proposed by various oil and gas service companies. Around that same time, places in the US such as the Barnett shale, Eagle Ford shale, and Marcellus shale began to develop. This would soon reshape not only the natural gas industry but the electricity industry as a whole. Suddenly, the prospects of a future of expensive

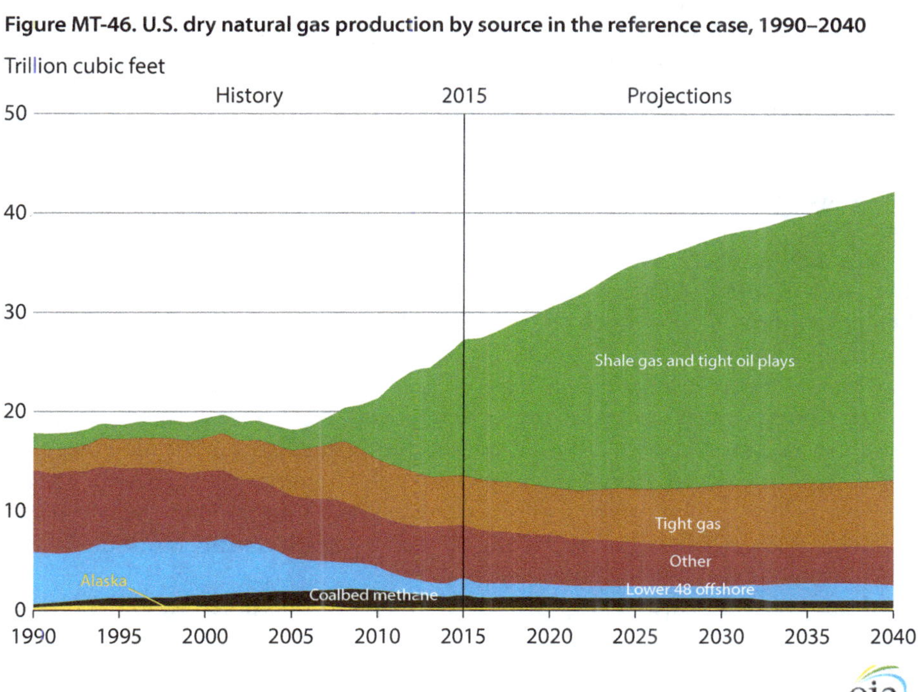

Fig. 4.4 Natural gas production over time and share of shale gas, tight gas, and other unconventional sources

natural gas flipped to a future of cheap natural gas. The US ended up building LNG export terminals to participate in the growing natural gas trade movements, involving LNG, which are led by Qatar and flows to southeast Asia.

But these new developments in extraction techniques did not come without controversy. The hydraulic fracturing or fracking process begins with permitting. Permits to drill wells are generally required by state regulatory agencies (or federal ones on public lands) before construction can begin. State laws and regulations for well permits differ. For example, setbacks from structures or sensitive landscape features vary among states, and data collection and reporting requirements are also different. The next stage is site development. If a well is being developed at an undeveloped site, a "pad" must be constructed before drilling can begin. Pads provide a level working surface free of obstructions. As enabled by horizontal drilling methods, multiple wells or horizontal legs may be advanced from a single pad. Horizontal legs can sometimes be two miles or more underground, and perhaps deeper in the future. In addition to the well pad, site development may be required for areas associated with pipelines or other supporting infrastructure. Issues related to pad construction include road making and, where applicable, deforestation.

The next phase of a fracking job is water acquisition. The fracking process involves injecting large volumes of water at high pressure dosed with a small volume of chemicals into the Earth. When water is injected into wells, a chemical mixture is added to prevent corrosion and algae formation and ultimately to increase the quantity of hydrocarbons that can be extracted. This pressure and volume ultimately create a large network of fractures in the shale, later, releasing oil, natural gas, or both. Water is acquired via direct extraction by the operator from surface or groundwater, by purchasing water, or by reusing fluid from previously drilled and fractured Marcellus sites, or some combination. Reuse of fluid is becoming increasingly common as companies receive scrutiny for freshwater use. Water is mixed with chemicals such as corrosion inhibitors, surfactants, and biocides.

Drilling commences once proper permits are approved and the site is prepared. The start of the drilling process is sometimes referred to as "spudding" a well. Unconventional shale wells begin with the drilling of a vertical borehole. The vertical borehole segments are then lined with a casing composed of steel that is sealed into place with cement down to a depth of up to 100 feet. Properly installed, casing protects groundwater, and it is common to extend the surface casing several hundred feet below any aquifers. With a surface casing in place, the borehole begins to curve, and an intermediate casing to prevent fluid migration and prevent borehole collapse is installed up to a point known as the kickoff. From here, the borehole is drilled laterally to run within and along the shale. Another casing is installed all the way through the end of the borehole.

After the well is drilled, explosive charges are set off to perforate the production casing. This puts the production casing in contact with the shale production zone. After a cleaning with hydrofluoric acid, the well is hydraulically fractured with a slurry of water, sand, and chemicals by injecting it under pressure. The sand or the other proppant material holds the fractures open after pressure is released. The fracking process involves blasting chemical-infused water at high pressure into the shale layer beneath the earth, which fractures open the rock, releasing the methane inside, a process known as completion. The massive industrial endeavor demands 2–4million gallons of water for a single lateral, and 15,000 to 60,000 gallons of chemicals (Hansen et al. 2013). Up to 600 different chemicals are contained in fracking fluid; some are unknown because they are considered proprietary. Produced water flows to the well surface, together with the oil and gas and is typically high in total dissolved solids (TDS) (including sodium, potassium, bromide, calcium, fluoride, nitrate, phosphate, chloride, sulfate, and magnesium), oil and other hydrocarbons, and "naturally occurring radioactive materials" (Chittick and Srebotnjak 2017, 503). Once the fracking process is completed, fluid that returns to the surface is referred to as "flowback" and is directed to temporary on-site storage where it is ultimately disposed of or reused. This flowback water returns to the surface within a week or so, but it can continue for months (American Petroleum Institute 2010). In addition to flowback fluid, naturally present brine fluid also returns to the surface.

Once a well begins producing gas, it can continue to produce for many years, although annual production volumes decrease rapidly. When production drops off, wells often need to be restimulated to increase production again, which requires another flush of several million gallons of water to increase production.

The history of extracting energy from shale goes back to the 1800s. Even though proponents have argued we are entering a new shale gas "revolution," production techniques like this have a long history. In 1821, decades before the first oil well was drilled, commercial shale gas was extracted in Fredonia, New York (Sovacool, 2014, 250). Traditional hydraulic fracturing has been in use since the 1940s. This involves drilling a vertical well shaft until it hits a layer of shale and then blasting the rock with chemically treated water and sand to break the rock and release the oil or natural gas inside. However, horizontal drilling became commercially viable starting around the mid-1990s. Horizontal drilling allowed access to a lot more shale reserves and largely contributed to the rise of fracking technology recently. In short, horizontal drilling does not require a vertical well shaft. A horizontal drill can follow the irregular shape of underground shale formations, opening up access to areas that were impossible to reach using the old system. The phenomenon of fracking in recent years, sometimes referred to as the "shale gas revolution," has only added to public concern about harvesting this form of energy. The fracking process has changed dramatically since it began, becoming more technologically cumbersome and complex as well as physically enormous (▫ Fig. 4.5).

Hydraulic fracturing or fracking is simultaneously touted as integral to America's energy future and could have severe implications for the local communities. The advent of horizontal drilling has opened up more resources that were previously unattainable, and over the next decade, it is believed that shale gas will be more than half of all natural gas supplied from the US. The problem is not whether enough natural gas resources exist on earth, it is whether human beings can access them and whether they should. For many years, shale gas was believed to be uneconomical. But with several key investments and innovations, first at the Department of Energy, then by the private sector, these concerns about economic viability went away.

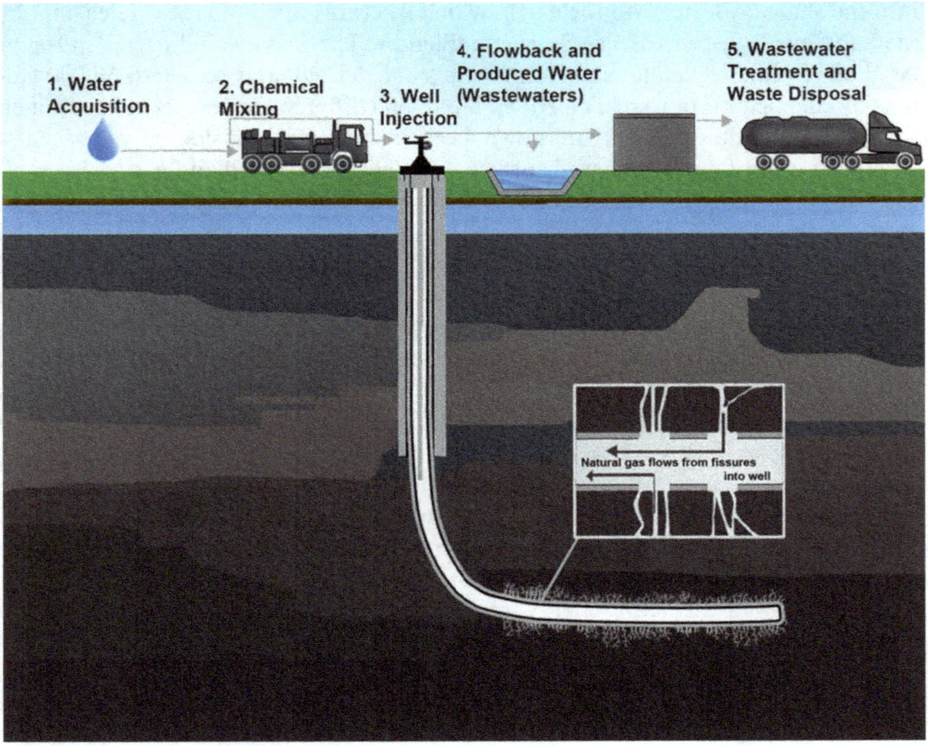

Fig. 4.5 Diagram of the hydraulic fracturing or "fracking" process

Fracking is a highly debated issue made even more challenging to sift through by corporate and political motivations, scientific uncertainty, and also hyperbole from environmental organizations. While there are benefits to fracking, there are also risks, which means the search for alternatives is also an important piece of the fracking debate. In the 2010 HBO film *Gasland*, narrator and filmmaker Josh Fox begins the movie at a family home, a lush green area of rural Pennsylvania in the Delaware River watershed. He receives a letter from a natural gas company that offers his family the option to lease their land to drill gas for US$100,000, no small sum for the Fox family. Throughout the remainder of the film, Fox drives through areas impacted by natural gas drilling that he calls "moonscapes"—from Pennsylvania to Texas to Colorado to Wyoming. The landscapes are fragmented by roads and dotted with well pads, leaving a moonscape-like pattern. The film paints a bleak picture of the impacts of natural gas drilling: water and air polluted, rivers and landscape degraded. Researchers in the film were able to make a simple plastic polymer from the chemicals coming off of condensate, a component of wet natural gas that comprises lighter hydrocarbons like propane, butane, pentane, and hexane, which eventually form smog when exposed to ultraviolet light.

Some researchers suggest that fracking for natural gas can put groundwater in and around the well site at risk. Impacts to groundwater quality include methane migration, shale formation brine and associated metals migration, and contamination of groundwater from drilling or flowback fluids, but the US EPA suggested that none of these are systemic, but rather are the result of poor or negligent work (US EPA 2015). Shallow groundwater is the most significant issue, given that that is close to the surface is often relied upon as drinking water, while deeper groundwater usually is not. There are many questions about the hydraulic connectivity between deep and shallow groundwater (Rabinowitz et al. 2015). Contamination of shallow groundwater from oil and gas development is through poor on-site management of flowback and drilling fluids (Holloway and Rudd 2013). Some have suggested that these poor management practices coupled with the scientific uncertainty warrant stopping fracking in practice (Howarth and Ingraffea 2011). Poorly managed and sited waste pits can also impair water quality via mechanical failure, such as when liners rupture and create direct contact with groundwater (Folger et al. 2012). Heavy metals such as barium and arsenic can impair groundwater if a waste pit liner fails or if spills occur outside the lined area. Buried waste such as drilling cores also poses water quality risks. Obviously, intentional dumping—which has been found to happen—will impact water quality (US EPA 2015). Though it has not been documented with certainty to date, damage to wells that occurs close to the surface might lead to the direct contamination of groundwater, as fracking or flowback fluid can be introduced to the subsurface, and possibly to groundwater resources, from the site of well failure. In the US, the Bentsen Amendment (1980) relaxes hazardous waste laws for drilling wastes.

Similarly, shale gas extraction can impact surface water by above-ground land management and off-site waste treatment. Spills, leaks, and other releases of flowback fluid can directly impair surface water because these fluids contain low-level radiation, heavy metals, and other contaminants. Surface water can be impacted by the permitted release of treated shale gas waste, including impacts of land-use change and the release of treated shale gas waste on total suspended solids (TSS) and chloride concentrations (Olmstead et al. 2013). Increased TSS concentrations would occur from runoff associated with well pads, pipelines, and roads The source of chloride is brine and flowback fluid. They found that chloride concentrations did impact downstream water quality, but they were not able to detect increased TSS concentrations (Olmstead et al. 2013). Numerous other studies are ongoing or underway.

Contrary to the viewpoint shown in the widely viewed film *Gasland*, in 2015 the EPA conducted an environmental assessment of the impacts of fracking on drinking water. The EPA released their fracking risk assessment back in 2015 that compiles up-to-date data and over 950 various sources of scientific reports (Gutin 2015). They found that while fracking does cause water contamination and pose a risk, the risk is not widespread. The study was extensive, but there were some major limitations: there was little data about water quality before fracking began at the sites they tested, and they were not able to test underground injection control (UIC) wells. The majority of the wastewater, approximately 98%, is disposed of using UIC wells. This is significant because UIC wells are responsible for most wastewater disposal related to fracking. Lastly, wastewater generated from the wells are managed by the EPA's permitted UIC wells, and studying them was outside of the EPA's

Fig. 4.6 A natural gas field in Jonah, Wyoming

research scope, but they found that structural failures in the UIC wells are common as well. In addition, since the study only tested water and soil contamination, it did not tackle the issue of air pollution surrounding natural gas drilling. While the EPA added a necessary scientific voice to temper the natural gas debate, due to the study's limitations, it certainly did not end it (Fig. 4.6).

If wells are not cased properly with cement, there is the potential for fracking fluids and other elements to seep into the ground and find its way to aquifers. The EPA confirmed that there have been instances of extreme spill rates of 12.2% due to well casing deficiencies (Gutin 2015). While there are very little data to assess the impact of UIC, it has been linked to an increase in earthquakes in states like Oklahoma where this practice is widespread according to the US Geological Survey (USGS). The EPA confirmed that there are significant impacts to drinking water, caused by hydraulic fracturing at select wells. However, with the limited amount of data available and the small scope of the study, the EPA could not conclude that there were any apparent public health risks inherent with the process of fracking (Gutin 2015).

There are pros and cons to natural gas development using shale resources. The public debate on fracking is ongoing and often shaped by partisan beliefs, in the absence of a large knowledge base with better and more objective facts (Christenson et al. 2017). The debate on fracking is characterized by economic positives weighed against quality-of-life negatives. Hydraulic fracturing has stimulated natural gas production, which has been stagnating since the 1990s. Other benefits include job creation and a dramatic increase in the amount of natural gas that the US can supply (Christenson et al. 2017, 407). This lessens US dependence on foreign suppliers

Fig. 4.7 Fracking has drawn widespread protest and environmental concern related to production processes and climate

and on coal as an energy source. Natural gas is also much less energy intense than coal and thus a better alternative for the atmosphere; an atmosphere that only has a limited capacity to absorb more GHGs without the earth warming to dangerous levels. A Massachusetts Institute of Technology study found when forecasting different electricity scenarios that the scenario with accelerated shale gas production would reduce US emissions by about 1/5th compared to current conditions (Sovacool 2014) (Fig. 4.7).

One important consideration for not only fracking but all natural gas and associated infrastructure is the role of the potent GHG methane emissions in adding to the overall impact of natural gas. It is methane migration that causes the "flaming faucets" seen in the documentary *Gasland*. But it is probably more important to recognize the impacts on GHGs as the escape of fugitive and other methane emissions could undermine the GHG savings associated with displacing coal with natural gas. One study from 2011 suggested that methane emissions on the 20-year GHG horizon erode any savings associated with displacing coal (Howarth et al. 2011). A study in 2018 found that the emissions of methane amounted to about 2.3% of total US production (Alverez et al. 2018).

One way that oil and gas extraction, production, and refining industries have dealt with methane emissions historically is via flaring. Companies simply burn off methane where it is uneconomical in oil production and in refineries. This reduces the GHG impact proportional to the global warming potential of CO_2 versus CH_4 and also causes air pollution, especially when flaring involves wet natural gas. Oil refineries often flare natural gas because it remains in some of the petroleum they refine, or it is produced by its processes. In countries like Nigeria, where there is significant oil extraction, natural gas is flared instead of being captured and delivered to customers. It is estimated that 2.5 billion cubic feet/year = ~40% of all of African energy consumption (World Bank 2019).

Despite all the challenges of fracking, there are benefits as well. Local and regional benefits from the extraction of shale gas and natural gas include economic development—employment, jobs, infrastructure, revenues, and taxes, all promise to reward communities with increased economic activity and employment opportunities. As was mentioned earlier, fracking led to economic expansion and job creation after the recession in 2009. But the economic downturn in 2020 showed the industry in a very vulnerable position economically.

Increased natural production from fracking also has lowered the price of natural gas. In 2006, natural gas prices were US$8 per 1,000 cubic feet (cf or ft^3); by 2012, the price was down to US$2.89 (Turcotte et al. 2014). This may be a temporary price depression as there are many complex factors that contribute to the cost of natural gas. Because the industry brings economic growth to areas of the country that need it, there are many communities that support the practice. For example, in 2012, Pennsylvanians earned US$1.2 billion in royalties from new wells. That is why there are several studies that show that approval or disapproval of fracking activities changes as a function of distance from wellhead and whether one gets royalties or not.

With all these benefits and drawbacks, one thing is clear: if the US is going to continue using natural gas, the process needs to be more sustainable because natural gas is a limited resource. There are several avenues for reducing carbon emissions, including reducing demand and switching to renewable energy sources. One major way to reduce carbon emissions from our energy sources is decarbonizing the supply by blending with renewable biogas from wastewater treatment facilities or dairies.

Biogas is a promising way to decarbonize natural gas. Biogas, created through a biological process called anaerobic digestion, consists mainly of methane and carbon dioxide. Biogas cleaning, or upgrading, either removes the CO_2 or converts it to CH_4. While five processes of biogas creation are commercially available, more are under development, though the available supplies may be regionally constrained. This area of research is a promising alternative, or supplement, to the US shale reserves and their accompanying environmental and health risks. Biofuels, out of current sources of alternative energy for liquid fuels, show the most promise of replacing or at least displacing some fossil fuels; by 2040, biofuels will account for 2% of US total energy needs.

The big four shale plays for natural gas in the US are the Marcellus, Haynesville, Fayetteville, and Barnett shale. Our study in 2013 found that 4.3 million gallons of freshwater were injected into the ground per fractured well in Pennsylvania surveyed from 2008 through 2013. Of those 4.3 million gallons, only approximately 6% of the injected fluid is recaptured and returned to the hydrologic cycle (Hansen et al. 2013). Ostensibly, these water use numbers are getting better as more reused water is used, and techniques become more water efficient. The key takeaway is that there are considerable impacts if the freshwater used is taken from a place under water stress.

To make natural gas extraction more sustainable, the International Energy Agency (IEA) proposed industry best practices. IEA's proposal sums up the measures drilling companies should take in order to responsibly and sustainably extract shale gas. The best practices include taking into account local communities, treating water responsibly, and eliminating leaks and emissions. Adoption of best practices could increase the cost of a well, but decrease the amount of water use and air emissions.

Another set of considerations posed by hydraulic fracturing are earthquakes. Many scientists believe there is a link between wastewater disposal, which is mostly from fracking operations and earthquakes. Research from Oklahoma shows that the disposal of wastewater is linked to earthquakes (US Geological Survey (USGS) 2019). Certain parts of Oklahoma are now at the same risk of earthquake—like California. Similar findings have been made in California, where injection wells are used (Ellsworth 2013). This is due to forcing high-pressure water beneath Earth's surface, which can shift the natural tectonic plates. Scientists have noticed an increase in mini-earthquakes that occurred as new oil production took place in Southern California.

Methane is also found as natural gas hydrates or hydro-methane in ocean deposits or permafrost. Gas hydrates are natural gas trapped inside a lattice of ice, usually trapped under a layer of sedimentary rock. Some gas hydrates are found in deposits below frozen tundra permafrost. In some climate change feedback scenarios, these deep-sea and tundra hydrates and the potent GHGs that go with it bubble to the surface. Some companies have explored extracting methane hydrates for commercial gas production. Some of this natural gas are biogenic in nature, though often sites are found in fossil fuel deposits. There may be more methane trapped in hydrates than any other deposit or formation on Earth. The USGS operates a Gas Hydrates Laboratory to conduct science on this not-well-known natural resource (USGS 2018). Some energy analysts think that oceanic methane gas hydrates could upend the natural gas markets once again (Gilbert et al. 2020).

4.4 Petroleum

Petroleum or crude oil is the natural resource that is used to produce several final products. The most familiar to readers will be gasoline and diesel. Readers from Europe may refer to some of these final products as "petro"—but petroleum refers to the primary energy resource extracted from the Earth. Other end products of petroleum, in addition to motor gasoline and diesel, include distillate fuel oil, kerosene, liquefied petroleum gas, petroleum coke, still gas, and residual fuel oil. Each end product has a different emissions factor for environmental emissions to account for differences in refining processes and amounts of primary energy used to produce the end product. During the early years of petroleum extracted, the primary end product sold was kerosene—the product that much of the early Rockefeller fortune was built upon, as the liquid fuel was widely used to light lamps in place of whale oil blubber and ethanol (Fig. 4.8).

By 1970, crude oil supplied nearly half of the world's energy, 43% in the US, and 50% in the EU (Smil 2012). Global oil consumption is 76 million barrels per day.

Fig. 4.8 Petroleum refining in Anacortes, Washington

Despite strong climate action and renewable energy adoption, oil consumption is expected to increase to 123 million barrels per day by the year 2025, although those trends may already be outdated with the growth in electric vehicles. In the 1970s, during and after several oil crises caused price spikes and shortages, there were pushes to transition away from oil on the grounds of energy security. The flows of oil were dictated by Organization of the Petroleum Exporting Countries (OPEC), and there were concerns that foreign policy was beginning to be shaped by energy policy. There were also suggestions that the US dependence on petroleum led to oil wars—the so-called blood-for-oil concept.

Another early motivation for transitioning away from oil and petroleum was a concern that the supplies were running out. By the early 1970s, Texas peaked production, then steadily declined. A few years later, the US too began to slide back. It was not until the advent of horizontal drilling and hydraulic fracturing that these two areas recovered. The "peak oil" or "Hubbert's peak" hypothesis argued that beyond some point, the cost of recovering petroleum would escalate in price as the cost of extraction went up (Inman 2016). This would push oil into a period of terminal decline. Today, most geologists do not believe there will be a shortage of petroleum available for humans anywhere in the near term. But the costs of obtaining oil from the deep seas, or remote lands, for example, are expensive. The rate of new oil and gas field discoveries peaked in the 1960s. Recent increases in oil production in the US are due to the shifting of unviable shale resources to viable ones since the mid-2000s.

While often described as a homogenous category, petroleum comes in different types with different physical properties. Two important dimensions are light versus heavy, and sweet versus sour. Light crude oil is less viscous and easier to work with at refineries. The sweet or sourness of crude oil is a measure of the sulfur content of petroleum. The light-sweet combination is the most desirable for refineries. Sour crude can be corrosive to work with and often more energy-intensive to process. Petroleum is also sold through different market channels; West Texas Intermediate (WTI) and Brent are described as benchmarks for prices and listed in most financial sections of newspapers and online commodity indices. WTI is also known as Texas light, or, to some others, it is called sweet crude. Brent crude is less sweet and heavier than WTI but also considered on the light, sweet spectrum. Dubai crude and Urals crude are benchmarks for sour, heavy oil. These sour and heavy types of petroleum are more costly to process. In some refineries, high sulfur content can corrode the equipment if the sulfur is not removed or equipment not upgraded. If some sweet light crude production comes off-line for some reason (repair, accident, civil conflict), gasoline prices will often go up, as there is more energy demand to supply gasoline.

Conventional petroleum production is a simple process where a well is drilled, a well-pad is constructed, and eventually, auxiliary tanks, pumps, and equipment are added. Natural gas is commonly co-produced with petroleum, and that methane must be either stored or flared off. This flaring can cause localized air pollution. Oil fields decline in output by about 3% a year, until, at some point, it becomes too expensive to extract the final amounts without further stimulation with water or carbon emissions.

Unconventional techniques used to recover oil include enhanced oil recovery (EOR) and fracking techniques similar to natural gas production. The Bakken shale of South Dakota in the US is a resource made economic by fracking and has led to a large boom and bust cycle in oil production. It was believed that the shale oil would not be obtainable from the shale oil rock. The question is what happens if oil prices fall when scarcity sets in? Will it hasten transitions toward more viable resources or will it be difficult to promote exploration and development of oil wells when they might cost more to develop when they are worth?

Why are some countries so dependent on petroleum? The US uses so much oil/petroleum compared to other nations because it lacks high-speed rail, found throughout Europe and Japan—using significantly less energy to move people regionally. Second, widespread use of diesel vehicles in Europe come with higher efficiency combustion engines, meaning that there is less fuel required to power cars there because less is wasted as heat. Diesel engines convert upwards of 35% of the fuel into motion, whereas gasoline-powered internal combustion engines are more around 20%. Third, inaction on fuel economy standards (in the US corporate average fuel economy (CAFE) standards) lead to stagnation in the efficiency per mile of automobile manufacturer fleets, leading to very little improvement in a vehicle's fuel performance.

Oil is also a political actor onto itself. The received wisdom is that the dependency of an economy on cheap oil stems from cabals of powerful political actors, or collective imperatives of imperialism, or even rampant conspicuous consumption. In these dominant narratives, oil has agency, invoking wealth and poverty, driving booms and busts, and having implications for freedom and authoritarianism. Geographer Matthew Huber (2013) offers a framework that offers a way to think about the deep cultural inscriptions of oil on the American way of life that deepen its dependency. Huber argues that the dominant narratives of US oil addiction fetishize oil as a thing. Instead, he sees oil as a socio-ecological relation, drawing on Marx's productive forces, Gramsci's view on hegemony, and Foucault's micro-politics and the production of subjectivities. Using this framework, Huber shows how the embodied practices through which subjects come to experience oil in daily life are acts of social reproduction—the social practices necessary to bring workers back to work. "Without oil, life, family, and everything else could not be reproduced" (Huber 2013). This individualistic notion ties oil prices to a form of cheap energy populism, often one with very authoritarian tendencies. Because higher prices interfere with social reproduction, they undermine the freedom and choice needed to improve everyday life. We see this evident in the culture wars today. Through the 1980s and 1990s, the concerns about running out of petroleum subsided. There was plenty of oil. It was just increasingly expensive to extract, produce, and deliver. But by the 2000s, high prices for energy were again at the center of US politics. This time much of the clamor about high-energy prices ran contrary to emerging concerns about "blood for oil" and climate change.

One pivotal player in global oil production is the Organization of the Petroleum Exporting Countries (OPEC). The cartel operates by ensuring that prices and production outputs are set to achieve favorable outcomes for members. Saudi Arabia is what oil economy experts call a swing producer. This means they can swing production in short notice because of significant amount of oil production capacity. But recent years have seen some diminishment in the power of OPEC and the rise of the US and natural gas industries over petroleum ones. The Permian basin in the Texas–New Mexico region is now considered potentially the largest oil field in the world.

As new petroleum resources are sought further afield, new risks will be presented. The Deepwater Horizon spill released over 4 billion barrels of oil into the Gulf of Mexico, impacting tourism, fishing, and other dependent activities. Oil clean-up is a multi-step process, involving wildlife rescue and containment first. Oil is collected with contaminated water by skimming the surface. Also sometimes used is "in situ burning," a process where oil is burned on the water. Finally, chemical dispersants can be released to break down large areas of oil, as was the case with the BP spill, but there is some evidence that this caused problems elsewhere in the water column.

In coastal US states like California and Florida, new drilling is not allowed within three miles offshore. In California, the State Lands Commission has jurisdiction over tidelands and waters extending roughly three miles offshore. Currently, there are 23 oil platforms in federal waters off California and 4 in state waters, meaning that most facilities fall under federal management. California is a state with a very diverse geographical background, where both offshore and onshore fracking occurs. The state has a long history of oil operation that started in the mid-nineteenth century with the earliest report of fracking in 1953. The Santa Barbara channel is said to be one of the most ideal areas for offshore fracking. Santa Barbara contains an abundance of fossil fuel along with super-rich biological diversity oil sources. The southern coast of California hosted one of the world's first offshore platforms at the Summerland Oil Field.

With Texas in the lead for oil production in the US, California ranks third along with North Dakota in 2020. The introduction of hydraulic fracturing has helped restore hundreds of old oil and natural gas wells, adding to the tens of thousands of active oil and natural gas wells today. California oil is viscous and can be produced by steam injection, a particularly energy-intensive extraction technique. In 2015, the oil and gas production contributed about 2.7% of California's GDP—over US$50 billion—and supported about 368,100 jobs. The Monterey Formation, located in Southern California, contains nearly 24 billion barrels of oil. With the technological advancement of hydraulic fracturing and vertical drilling, oil may be more easily extracted from the Monterey Formation.

Globally, there are a handful of major players in offshore oil production. Norway successfully utilizes its oil wealth to support social services. Elsewhere, countries like Angola and Nigeria find less benefit from living among oil infrastructures. As pressure to develop these resources in international waters grows, it will be critical that these pursuits abide within the framework set out by the United Nations framework called the Law of the Sea.

The 1969 Union Oil spill from a platform near Santa Barbara, California, catalyzed the public to have a mostly negative view of the offshore oil and gas industry (Clarke and Hemphill 2002). The spill that occurred in 1969 remains the most massive oil spill in California's history and ranks third in the US. Union Oil company's platform was six miles off the Southern California coastline and suffered a massive blowout. About 220,000 gallons of oil spilled into the Pacific Ocean, covering over 800 square miles of sea. The oil slick traveled from the platform in Southern California to the four northern Channel Islands. It became a major headline for weeks at numerous media outlets and was widespread throughout radio talk shows and televisions around the world.

Many point to the spill as catalyzing the US federal government to set up the Environmental Protection Agency (EPA) to help prevent future oil spills. They also passed the Clean Air Act and Clean Water Act. After years of legal actions, Santa Barbara was able to compromise with oil companies that wouldn't allow drilling within a certain distance to the city of Santa Barbara.

From an economic perspective, supporters of fracking present the more optimistic view. Data suggest that in 2008 alone, Pennsylvania brought in US$238 million in tax revenues and created 29,000 new jobs due to shale gas extraction and supporting industries. In 2011, the Barnett Shale in Texas yielded US$11.1 billion in annual revenues and created over 100,000 jobs, nearly 10% of regional employment (Sovacool 2014, 254). Natural gas prices in the US have fallen dramatically from US$13 per million BTUs to US$1–2 in 2012 due to shale gas, and cheap gas also creates cheap electricity (Sovacool 2014, 253). These benefits from low energy costs have an economic value that is hard to quantify, but must have some effect.

4.5 Tar Sands, Oil Sands

Some call this natural resource "tar sands," while others call it "oil sands." The oil and gas industry prefers oil sands to emphasize the resource, whereas most environmental organizations refer to it as tar sands because it conveys some of the properties of the viscous material that can be more challenging to clean up. Oil sands are a combination of sand, clay, water, and bitumen. Bitumen is a kind of hydrocarbon and can be used for any products of oil and for gasoline. Nearly 75% of tar sands on the Earth are in Alberta, Canada. Producing and processing tar sands oil creates roughly 14% more greenhouse gas emissions than regular conventional oil. There are many controversies over using tar sands as a main energy source since it creates high amounts of carbon dioxide emissions.

The Keystone Pipelines takes oil from Canada to the US in four phases. The first phase takes petroleum from Alberta, Canada, through the Dakotas, Nebraska, and ends in Illinois. The second phase of the pipeline starts at the connection in Nebraska and then ends in Cushing, Oklahoma, where large volumes of oil are stored before moving them to refineries in Texas along the third phase. First proposed in 2008, the fourth phase—The Keystone XL Pipeline—is more controversial. This phase connects Hardisty, Alberta, Canada, through Montana, South Dakota, and eventually the rest of the system in Steele City, Nebraska. The purpose of the pipeline according to project developer TransCanada is to provide additional oil supply and create much-needed jobs for Americans, and increase tax revenues.

Tar sands is at the center of the controversy over the Keystone XL pipeline. For every gallon of gasoline produced by tar sands, about 5 gallons of freshwater are consumed during the extraction, upgrading, and refining process (Barbose, 2013). When tar sands are mined on the surface, wastewater is stored in evaporation ponds.

One area of contestation is that the pipeline would run over sandhills in Nebraska, which are interconnected by the Ogallala Aquifer. Many farmers do not want this pipeline because they worry how a tar sands spill might impact water quality, which is critical to the farm economy there. A leak could contaminate the aquifer that so many depend on. In summer 2020, the Keystone XL was dealt a major legal setback.

Another controversial pipeline in recent years is the ~1200-mile pipeline from North Dakota to Illinois refineries called the Dakota Access line. In this case, in addition to environmental and climate action groups, many indigenous persons opposed the Dakota Access pipeline for reasons related to not just water quality but cultural resources, prior informed consent, and injustices related to a settler-colonial type (Powys-Whyte 2017). This frame allows for a decentering of procedural injustices, or disagreements over the minutia of specific project elements, and to zoom out and see this as part of a larger colonial history of broken treaties, land grabs, and systematic racism. Much like the tar sands pipeline, the Dakota Access line was ordered to shut down, dealing a hard-fought victory to indigenous activists and their climate justice collaborators.

The story of tar sands will continue to be that it becomes viable as a source for making gasoline, at higher prices, but that it will always be a water- and greenhouse gas–intensive primary energy source. Its future is further clouded by the challenges associated with moving it around. Whether by pipeline or train, it is a pretty risky proposition, and few will welcome infrastructures activists nickname "bomb trains."

4.6 Oil Shale

Oil shale is a viscous energetic fluid sometimes called kerogen oil. It is a different natural resource than the previously described shale oil, like that found in the Bakken. Oil shale is mostly a niche or pilot or experimental extraction technology in a few places on the planet. The US supply of oil shale is mostly found in Wyoming, Utah, and Montana, and this represents most of the oil shale on the planet, although some oil shale is found in Russia, Brazil, and the Democratic Republic of Congo.

The process most commonly used to extract oil shale is called retorting, which essentially boils water to extract the fuels in the mixtures of water, kerogen, clay, and sand. This means the extraction process requires large amounts of water, which is especially problematic in the dry areas where it occurs. Oil shale attracted a lot of research and development support during the energy crises of the 1970s. It is very unlikely this technology will be developed for gasoline, given the direction of climate policy and high cost of development. Yet, shale gas once was thought to be uneconomical. Now it's a leading primary source of energy.

References

Alvarez, R. A., Zavala-Araiza, D., Lyon, D. R., Allen, D. T., Barkley, Z. R., Brandt, A. R., & Kort, E. A. (2018). Assessment of methane emissions from the US oil and gas supply chain. *Science, 361*(6398), 186–188.

American Petroleum Institute. (2010). Water management associated with hydraulic fracturing. Report. Washington, D.C.

Barbose, J. (2013). What are tar sands?. Union of Concerned Scientists. White paper. Available at: https://www.ucsusa.org/clean-vehicles/all-about-oil/what-are-tar-sands#.WwOcB0xFz3.

Chittick, E. A., & Srebotnjak, T. (2017). An analysis of chemicals and other constituents found in produced water from hydraulically fractured wells in California and the challenges for wastewater management. *Journal of Environmental Management, 204*, 502–509.

Christensen, I., Lange, I., Sowers, G., Abbud-Madrid, A., & Bazilian, M. (2019). New policies needed to advance space mining. *Issues in Science and Technology, 35*(2), 26–30.

Christenson, D. P., Goldfarb, J. L., & Kriner, D. L. (2017). Costs, benefits, and the malleability of public support for "Fracking". *Energy Policy, 105*, 407–417.

Clarke, K. C., & Hemphill, J. J. (2002). The Santa Barbara oil spill: A retrospective. *Yearbook of the Association of Pacific Coast Geographers, 64*, 157–162.

CLB (China Labor Bulletin). (2019). Deaths from coal mine accidents in China fall to new low of 333 in 2018. https://clb.org.hk/content/deaths-coal-mine-accidents-china-fall-new-low-333-2018

Ellsworth, W. L. (2013). Injection-induced earthquakes. *Science, 341*(6142), 1225942.

EPA. (2011). Regulatory impact analysis for the final mercury and air toxics standards. EPA-452/R-11-011.

EPA. (2015). EPA's study of hydraulic fracturing for oil and gas and its potential impact on drinking water resources. https://www.epa.gov/hfstudy.

Folger, P., Tiemann, M., & Bearden, D. M. (2012). The EPA draft report of groundwater contamination near Pavillion, Wyoming: Main findings and stakeholder responses. *Congressional Research Service, 42327*.

Gabrys, J. (2017). Citizen sensing, air pollution and fracking: From 'caring about your air' to speculative practices of evidencing harm. *The Sociological Review, 65*(2), 172–192.

Gilbert, A., Bazilian, M., & Loza, S. (2020, January 9). The world's next energy bonanza. *Foreign Affairs*. https://foreignpolicy.com/2020/01/09/fracking-oceanic-methane-hydrates-global-energy-landscape-bonanza/.

Gutin, O. (2015). EPA releases fracking risk assessment. Retrieved from http://www.eesi.org/articles/view/epa-releases-fracking-risk-assessment.

Hansen, E., Mulvaney, D., & Betcher, M. (2013). Water resource reporting and water footprint from Marcellus Shale development in West Virginia and Pennsylvania. *Downstream Strategies*. https://earthworks.org/publications/water_resource_reporting_and_water_footprint_from_marcellus_shale_developme/.

Holloway, M. D., & Rudd, O. (2013). *Fracking: The operations and environmental consequences of hydraulic fracturing*. New York: Wiley.

Howarth, R. W., & Ingraffea, A. (2011). Should fracking stop? Yes it's too high a risk. *Nature, 477*, 271–273.

Howarth, R. W., Santoro, R., & Ingraffea, A. (2011). Methane and the greenhouse-gas footprint of natural gas from shale formations. *Climatic Change, 106*(4), 679.

Huber, M. (2013). *Lifeblood: Oil, freedom, and the forces of capital*. Minneapolis: University of Minnesota Press.

Inman, M. (2016). *The oracle of oil: A maverick geologists quest for a sustainable future*. New York: W.W. Norton.

Mooney, C. (2011). The truth about fracking. *Scientific American*, 80–85. https://www.scientificamerican.com/article/the-truth-about-fracking/.

Mumford, L. (1934). *Technics and civilization*. Chicago: University of Chicago Press.

New York Times. (2009, January 6). Hundreds of coal ash dumps lack regulation. http://www.nytimes.com/2009/01/07/us/07sludge.html.

Olmstead, S. M., Muehlenbachs, L. A., Shih, J. S., Chu, Z., & Krupnick, A. J. (2013). Shale gas development impacts on surface water quality in Pennsylvania. *Proceedings of the National Academy of Sciences, 110*(13), 4962–4967.

Pai, S., Zerriffi, H., Jewell, J., & Pathak, J. (2020). Solar has greater techno-economic resource suitability than wind for replacing coal mining jobs. *Environmental Research Letters, 15*(3), 034065.

Rabinowitz, P. M., Slizovskiy, I. B., Lamers, V., Trufan, S. J., Holford, T. R., Dziura, J. D., et al. (2015). Proximity to natural gas wells and reported health status: Results of a household survey in Washington County, Pennsylvania. *Environmental Health Perspectives, 123*(1), 21.

Smil, V. (1991). General energetics: Energy in the biosphere and civilization. In *General energetics: Energy in the biosphere and civilization*. New York: Wiley.

Smil, V. (2012). A skeptic looks at alternative energy. *IEEE Spectrum, 49*(7). https://spectrum.ieee.org/energy/renewables/a-skeptic-looks-at-alternative-energy.

Sovacool, B. K. (2014). Cornucopia or curse? Reviewing the costs and benefits of shale gas hydraulic fracturing (fracking). *Renewable and Sustainable Energy Reviews, 37*, 249–264.

The Energy Transition Show. (2019). Episode 92. Financing Coal Retirements. https://xenetwork.org/ets/episodes/episode-92-financing-coal-plant-retirements/.

Turcotte, D. L., Moores, E. M., & Rundle, J. B. (2014). Super fracking. *Physics Today, 67*(8), 34–39. https://doi.org/10.1063/pt.3.2480.

Udall, M. K. (1978). The enactment of the surface mining control and reclamation act of 1977 in retrospect. *West Virginia Law Review, 81*, 553–557.

USGS (U.S. Geological Survey). (2018). Woods Hole Coastal and Marine Science Center, Gas Hydrates Project. https://woodshole.er.usgs.gov/project-pages/hydrates/.

USGS (U.S. Geological Survey). (2019). Oklahoma now has more earthquakes on a regular basis than California. Are they due to fracking? https://www.usgs.gov/faqs/oklahoma-now-has-more-earthquakes-a-regular-basis-california-are-they-due-fracking?qt-news_science_products=3#qt-news_science_products.

Wang, Q., & Li, R. (2016). Journey to burning half of global coal: Trajectory and drivers of China's coal use. *Renewable and Sustainable Energy Reviews, 58*, 341–346.

Whyte, K. (2017). The Dakota access pipeline, environmental injustice, and US colonialism. *Red Ink: An International Journal of Indigenous Literature, Arts, & Humanities, 19*(1), 154–169.

World Bank. (2019). Global gas flaring reduction partnership. https://www.worldbank.org/en/programs/gasflaringreduction.

World Energy Council. (2017). Coal mining. https://www.worldcoal.org/coal/coal-mining.

Energy and the Environment II: Nuclear and Renewables

Contents

D. Mulvaney, *Sustainable Energy Transitions*,
https://doi.org/10.1007/978-3-030-48912-0_5

Learning Goals

This chapter aims to improve readers' energy acumen and ability to:

- describe how different energy technologies work;
- assess the generation potential of different renewable and clean energy sources and technologies; and
- understand the social and environmental dimensions—the socio-ecologies—of existing and potentially widespread deployment of renewables and clean energy in the future.

Overview

This chapter describes energy resources that are low carbon and renewable. The fossil fuels mentioned in ▶ Chapter 4 could be made low carbon in some cases, but these here mostly avoid carbon altogether. In 2020, these technologies comprise 20–30% of global energy demand depending on how they are measured. Globally, power demand is on the order of 20 terawatts. Solar and wind energy combined passed a total of 1 terawatt installed in 2018 according to *Bloomberg New Energy Finance* (BNEF 2018), and the second terawatt will be here in 2023. Hydropower is the most widely deployed renewable energy resource in 2020, followed by wind and solar power.

Low-carbon energy sources like solar power, wind turbines, nuclear power, hydropower, geothermal, biofuels, and carbon capture fuels all play roles or are subjects of debate about decarbonization strategies for energy transition. The energy resources and technologies presented in this chapter, as solutions to the climate emergency, also cause socio-ecological transformations with impacts that should be evaluated and considered by decision-makers and the public. This chapter describes each low-carbon energy source and the technologies used to transform it into useable energy, key attributes, and socio-ecological impacts to allow for a comprehensive view on the energy choices that face human civilization.

5.1 Uranium and Nuclear Power

The hotly contested debate in the decarbonization strategies universe is what to do about nuclear power. Once the superstar of the energy family in the 1950s, the nuclear power industry in 2020 is saddled with many challenges. Many older operating nuclear power plants struggle to compete with newer lower-priced sources. Newer projects face construction challenges, leading to cost overruns and delays. Only limited funding is available to build new projects, so there are fewer entrepreneurs and financiers pursuing next-generation nuclear facilities and technologies. While nuclear power plants produce so much low-carbon electricity, they may offer an opportunity to satisfy a large portion of electricity demand with low-carbon electricity.

Climate change activism and advocacy is now a global social movement. There are multiple currents, but most share in common an aspiration to rapidly move global energy sources away from fossil fuels. The global nuclear power plant fleet is around 450 reactors producing around 11% of global electricity (World Nuclear Association 2019).

Many years of opposition by some environmental and anti-war organizations, a discourse linking nuclear power to nuclear winter post-nuclear war stories, accidents, and cost overruns, had sidelined nuclear power for many decades in the US, and slowed its ascendance in developing countries. Despite years of stagnation, 19% of all electricity generated in the US in 2018 was still from nuclear power, a percentage that continues to go down from a peak in the 1990s. Only two new nuclear plants in the US were ordered after 1978, despite continued adoption globally. Most US nuclear power plants are beyond their expected lifetime and a handful are up for retirement in the near future or have reported concern about economic viability. Concerns about losing these sources of electricity have prompted some energy policymakers to consider ways to make them more competitive or offer a bonus for low-carbon resources. If nuclear power plants are replaced with fossil fuel resources—coal, natural gas—the overall implications would raise greenhouse gas emissions. Some places like California are replacing nuclear power plants with renewable energy and energy storage. The question of nuclear power has no one-size-fits-all solution. Different countries are going to have to think through these questions in different ways and decide what path they want to take.

The low-carbon emissions led to a brief nuclear renaissance by the time the United Nations' Intergovernmental Panel on Climate Change (IPCC) was awarded the Nobel Prize in 2007. But by March 2011, the nuclear renaissance discussion paused as the Fukushima Daiichi Nuclear Power Plant in Japan underwent a meltdown from damages caused by a tsunami that occurred after a magnitude 9.0 earthquake. The disaster sparked international debates on the usages of nuclear power generation and the place that it holds. Countries such as China and the UK remained firm in their plans of increasing nuclear power, while others like Germany and Japan took the opposite approach and began reexamining their national policies on nuclear energy (Brumfiel and Cyranoski 2011); Germany had a policy to phase out nuclear much earlier that was abandoned because of cost. The Fukushima accident had similar consequences to accidents at Three Mile Island and Chernobyl, negatively changing the public opinion regarding nuclear energy. It did so globally, and locally, as many nuclear power plants in Japan faced social opposition during efforts to restart them (Visschers and Siegrist 2013) (▫ Fig. 5.1).

Nuclear energy is a result of splitting uranium 235, an isotope found in rock at about 2–3% of uranium, depending on the origin of the ore. Most of the balance is non-fissile uranium, U-238. Each split uranium 235 atom releases heat, atomic fragments, and one or two neutrons can cause other uranium 235 atoms to also split. A consistent rate of radioactive decay of uranium 235 occurs in the core of a nuclear plant. This heat is used to make steam, turn a turbine, and spin a generator. The reactor core is kept cool usually by circulating water in the tower. Passively cooled power plants, which are not yet commercially in operation, propose using gravity, density, or some other physical process to ensure that cores of reactors are kept cool without the need for an external energy source. Most nuclear power plants are either boiling water reactors or pressured water reactors. The major difference is in how the heat is transferred in the process of making steam.

Fig. 5.1 Protest in Berlin by farmers against the siting of a nuclear waste dump near the border with Poland

Nuclear Radiation Primer

Nuclear energy refers to the energy that is released when a nucleus of an atom breaks apart or disintegrates. Nuclear radiation measurements can be confusing, but they are important to understand as a tiny bit of radiation is probably alright, but too much radiation can have significant impacts on living organisms. Part of the confusing nature of radiation measurement is that different kinds of radiation—X-rays, gamma rays, and beta rays—have different implications for different materials. Below are the different ways we approach radiation measurements. Note that some metrics measure the amount of energy, while other metrics represent a degree of biological injury.

- Becquerel—1 nucleus disintegration per second (s)
- Curie—37,000,000,000 Becquerel (Beq)
- Rad—radiation dosage of 1 joule (J) per kilogram (kg)
- Rem—Roentger equivalent mean dosage in rads to cause some biological injury as one rad of X-Rays or gamma rays
- Sievert (Sv) = 100 rem = 0.001 milli-Sievert (mSv)
- X-rays and gamma rays: 1 rad = 1 rem = 10 mSv
- Neutrons: 1 rad = 5–20 rem = 50–200 mSv
- Alpha-rays (He_4): 1 rad = 20 rem = 200 mSv

There is little debate about the dangers of high-level radiation and health impacts. But there is less consensus on low-level radiation and impacts on public health.

Uranium mining is an important supply chain impact consideration for nuclear power often ignored in climate action conversations today. In the US, environmental justice impacts have occurred on Native American lands, including the Church Rock disaster on Navajo lands in 1979, where a tailing pond dam collapsed and caused the largest unintentional release of radioactive material in US history. What makes uranium mining for nuclear power complicated is that it shares the same supply chain as nuclear weapons. Even if there were no more nuclear power plants, there would still be uranium mined in some places because of the dependency for the metal for nuclear warheads. Incidentally, some people suggest we transition away from keeping nuclear warheads around too.

Uranium makes up only two parts per million of the Earth's crust. But a single pound of uranium contains as much energy as three million pounds of coal. Uranium is much more energy dense than other energy sources.

Mining impacts remain a key area of environmental damage. The top two producers of uranium ore include Canada and Australia. In the US, uranium is mined mainly across the Western states in sandstone deposits along with coal, such as Wyoming and the Four Corner regions. The US was the first to develop extensive uranium mining with many so-called yellowcake towns scattered across the American West and at the time of the Manhattan Project to develop nuclear weapons. UO_2 fuel pellets are assembled into rods commonly made of a zirconium alloy. The pellets heat the rods, which are the source of heat for the steam turbine. Some nuclear fuel rods incorporate reused uranium and plutonium in a mixed-oxide fuel, MOX. This fuel can be made from weapons-grade plutonium for nuclear disarmament. Fuel rods in a nuclear power plant need to be replaced about every other year on average. The waste fuel rods are transported for fuel recovery or stored on-site surrounded by water, which shields radiation. The concern with some nuclear accidents has been with the loss of this water during an accident.

The typical nuclear power plant generates on average about 20 metric tons of high-level radiation waste annually (Kessides 2010). The nuclear industry produces 2,300 metric tons each year and has created around 70,500 metric tons since its inception (Davis 2012). The US has no central repository for nuclear waste, though there have been plans to build one in the Yucca Mountains near the atomic test site an hour west of Las Vegas.

Nuclear power plants can be important for jobs and financial support in the local economy. It can be difficult to assess the financial and community costs associated with decommissioning the power plant (Ballard et al. 1991). What does decommissioning actually look like from the perspective of jobs, community revenues, or justice? As nuclear power comes with considerable construction, regulation, and decommissioning costs, it does not qualify as a renewable source of energy.

Most global growth in nuclear power is in China. Nuclear power plants in the US were built from 1970 to 1990. About 95% of all power plants operating in 2018 were opened in this time period, meaning the youngest nuclear power plants are already 30 years old. Many attribute the slowdown in nuclear power plant construction to several forces, including the environmental and anti-war movements. Nuclear accidents in the history of nuclear power, including accidents at Windscale (1957), Cheylabinsk (1957), Saint-Lavient (1969), Three Mile Island (1979), and Chernobyl

(1986), played important roles in shaping public perceptions of the risks of nuclear power. Considering the myriad of factors that shape utility commitments to nuclear energy in the US, it is clear that these narratives—that ascribe significant roles to the environmental movement, anti-war movement, and Three Mile Island in fostering opposition to nuclear power—give too much agency to those elements of the story. They are partial explanations at best; more a perfect storm of heterogeneous associations—a term used by Jonathan Murdoch (1997) to describe interactions across seemingly disconnected subjects across what others put in either the nature or society domains. The stall of nuclear power in the US is a more multi-faceted story than one of losing public relations because of contamination issues. Electric utilities saw flat demand for electricity, cost overruns on projects in progress led regulators to shield ratepayers, and cheap oil and natural gas at the time made it easier to justify.

There are some ideas promoted in a second generation of reactor technologies and fissile fuels. Small modular reactors are proposed as a solution to the challenges of cost overruns. The basic idea is to make nuclear reactors manufactured products where costs reductions are possible from economies of scale. Some proponents of nuclear power propose thorium reactors in place of uranium- and plutonium-based ones, as thorium is more abundant in Earth's crust than uranium, does not produce the same amounts of high-level waste, and poses lower weapons proliferation risks. However, to produce electricity from thorium, it is first converted to protactinium by bombarding it with neutrons; this protactinium decays to form uranium 233. Uranium 233 is not found in appreciable quantities in Earth's crust and was discovered during the Manhattan project; scientists say it has more attributes that make it even more suitable than uranium 235 for making weapons. Currently, India operates the only thorium-based nuclear power plant in the world, the 30 kW Kalpakkam Mini Reactor.

So what should countries with aging nuclear power fleets facing costly repairs do with aging nuclear power plants in the US? In Germany? Japan? France? These closures could lead to increased electricity greenhouse gas (GHG) emissions, particularly from natural gas. But are there opportunity costs? If there are new nuclear power plants built, and uranium supply chains continue to develop, what can be done to protect those that live alongside the commodity chains? Future power plants will need to be smaller and more flexible to operate.

5.2 Wind Power

The origin of wind power is mostly from the sun, as solar energy drives temperature differences. Earth's rotation also contributes minor amounts of energy to wind power. Power from wind is first harvested as mechanical energy and then converted to electric power. Going back hundreds of years, humans used wind turbines to grind seeds, pump water, and sail on the seas. For 70 years, wind turbines have been used to generate electricity.

Recent years have seen a dramatic increase in wind power for electricity. Global wind installations in 2020 stand at 600 GW. Much of the global growth in renewables over the past two decades have been from wind farms. Wind turbines are being installed at extremely fast rates, while the turbine technology continually

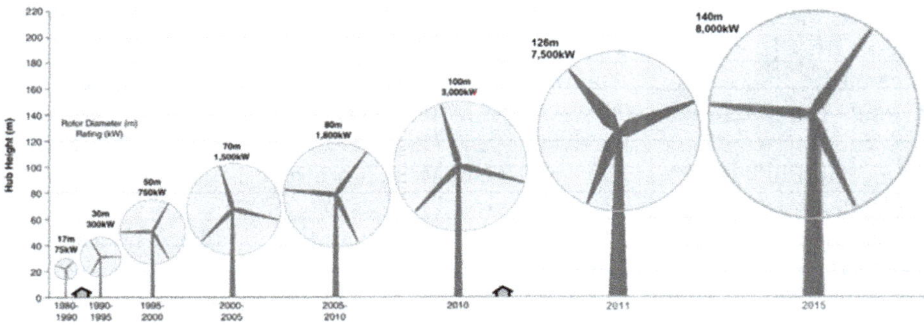

Fig. 5.2 The size of wind turbines continues to grow

improves. According to the Department of Energy, turbine capacity has increased dramatically since 2010. The average wind turbine capacity in 2020 is around 2.5 megawatts (MW), but the largest ones are getting huge (Fig. 5.2).

Definition

$$\text{Power} = 0.5 \times \rho \times C \times v^3 \times A_{\text{Sweep}}$$

V^3: meters per second—doubling of the wind speed results in an eightfold increase in power.

ρ: 1.2 kg/m^3 (kilograms per cubic meter) high-density air results in more power (altitude and temperature).

As: m^2 (square meters) area swept by blade; slight increase in blade length increases the area greatly.

C: efficiency coefficient, no units, different types of wind turbines have different maximum theoretical efficiencies (Betz limit $\approx$ 0.593 or 16/27^{ths}) but usually between 0.4 and 0.5.

Assignment 5.1 Wind Power Rises to the Cube of the Velocity

1. One location has wind that blows only 50% of the time at 10 m/s but is calm the rest of the time.
2. A second location has wind that blows all of the time at 5 m/s.
   - $P = 0.5\rho C_p v^3 A_S$
   - $\rho = 1.2$ kg/m^3
   - $A = 100$ m^2
   - C_p: 50% or 0.5

Both have exactly the same annual average wind speed. But how much electricity can each turbine produce over the course of one day? Estimate how much in each scenario. Look at the equation and write a few notes explaining why.

China is the global leader in wind power with 35% of the cumulative capacity for wind. Other leading countries as of 2018 include the US (17%), Germany (10%), Denmark (2%), India (6%), Brazil (2%), the UK (3%), Italy (2%), France (3%), Canada (2%), and Spain (4%) (Global Wind Energy Council 2018). Texas is the US leader in wind power, receiving over 17% of its electricity from wind in 2017. Some call the land from Texas and New Mexico north to Montana, Wisconsin, and Michigan the "Saudi Arabia of wind."

Wind power is strongest in the far northern and far southern hemispheres, where the westerlies dominate. Major obstacles to harnessing wind for electricity include the distance from the wind resources to population and load centers. Wind farms are starting to integrate battery storage, giving the intermittent energy source some ability to be dispatchable.

There is more power offshore than onshore because there are fewer obstacles to interrupt wind patterns. This attracts developers to offshore wind power, as the higher and steadier wind power in open seas means more electricity generation with the same turbine (Kaldellis et al. 2016). Onshore wind poses fewer challenges and costs less because of experience and ease of access. Offshore wind has been operating in northern Europe since the 1980s. The first offshore wind farm in the US opened in 2016 in Rhode Island, and more are planned on the Atlantic and Pacific coasts. Social science research from that Rhode Island wind farm shows the importance of community trust and engagement in the siting process (Firestone et al. 2020).

Floating wind technologies are required for deep-water deployment. Three types of floating systems include spar-buoy, semi-submersible, and the tension leg platform. Spar-buoy is stabilized by heavy material that extends deep and avoids damaging surface wave action. The semi-submersible floating turbine is for shallow water, up to 50 meters. The semi-submersible design is based on buoyancy and water-plane area to maintain static stability, but it is stable enough to support a wind turbine without mooring lines, and with a shallow draft that allows the structure to be towed fully assembled. Semi-submersibles allow assembly and maintenance to be performed at quayside rather than in the open ocean where more expensive vessels would be required (Musial et al. 2017). The tension leg platform uses a mooring-line tension. It can be difficult to deploy, but it is very stable once installed.

Building offshore wind farms could help regions with sufficient coastal access, land transmission constraints, and wind resource achieving the low-carbon goals. However, coastal and marine environmental impacts need to be identified and settled before engaging any projects that could impact the environment or communities.

Building offshore wind requires many miles of underground work to transport the cables, and this causes disturbance to benthic marine ecosystems and could even be deadly to humans (Kaldellis et al. 2016). Ecological impacts of wind farms are well documented, with leading impacts related to avian and bat species mortality, and also land-use change primarily from road development. Many of the earliest studies of avian collisions with wind farms came from the Altamont Pass wind projects in California. The high rates of bird collisions, with some very charismatic birds such as golden eagles, were widely cited. Many analysts attribute the high rates of mortality here to the location of the pass between the very important San Francisco Bay estuary, close to the Pacific coast, and the Pacific flyway that is in the Central Valley, further inland. This case points to the importance of location in understanding potential impacts. Also, in the past several years, this project has been largely repowered, where the main wind farms consisting of over 400 smaller turbines have been replaced by several dozen large-output turbines. The placement of these new turbines was partly informed by monitoring data, which is key to understand how to mitigate and minimize impacts to avian species.

5.3 Solar

The sun is the predominant source of energy, fueling biophysical activities on Earth. Wind, rain, life, and many other processes on Earth depend on inputs of solar energy. Of all resources available to humans, it is solar energy that is also the most abundant. The solar flux casts the planet in 1366 watts of power per square meter, or 2,500 exajoules of energy annually. Incoming energy from the sun is absorbed, reflected, or transmitted as it moves through the atmosphere, heating air and water, driving photosynthesis, affecting weather patterns, and adding energy to other biophysical processes. Enough solar energy hits Earth in one hour to power human civilization for an entire year, making it the least scarce of the energy sources available in the long term. Human civilization has always depended on solar energy as the sun powers crop photosynthesis, food webs, and the water and weather cycles (Crosby 2006). Fossil fuels are also solar energy captured and stored via photosynthesis and transformed by slow chemical processing for up to hundreds of millions of years. The relative ease of extracting very concentrated energies in hydropower, and fossil and fissile matter, led to the rapid displacement of the more diffuse renewable energies throughout the industrial revolution, right through the high-energy society of the Anthropocene.

Solar energy is collected for heating fluids (water and air) and generating electricity. Harvesting solar energy for electricity significantly lowers impacts to climate, air, and water compared to making electricity with fossil fuels.

To collect solar power, we need to make devices. Photovoltaic (PV) cell production requires raw materials extracted from mines, feedstocks produced at chemical plants, and fabrication facilities for manufacturing and assembling the semiconductor devices into modules. During installation, there can be impacts to ecosystems when utility-scale solar energy (USSE) projects are sited near wildlands or in sites that interfere with ecosystem processes (Lovich and Ennen 2011). At the same time, PV modules are the only renewable electricity source that can be readily integrated into the built environment and atop residential and commercial spaces and non-industrial zones.

Solar energy is also collected to provide space heating and to heat water. Hot-water heating, air-drying, and passive heating all involve harvesting energy from the sun to heat fluids, and they generally operate on the same basic principles. In many instances, the potential reduction in GHG emissions from solar heating can be higher than emissions saved with solar electricity because solar energy for heating homes passively or for heating water can displace distillate fuel oils and natural gas, whereas electricity displaced from solar power can be hydropower and other renewables, in addition to fossil fuels.

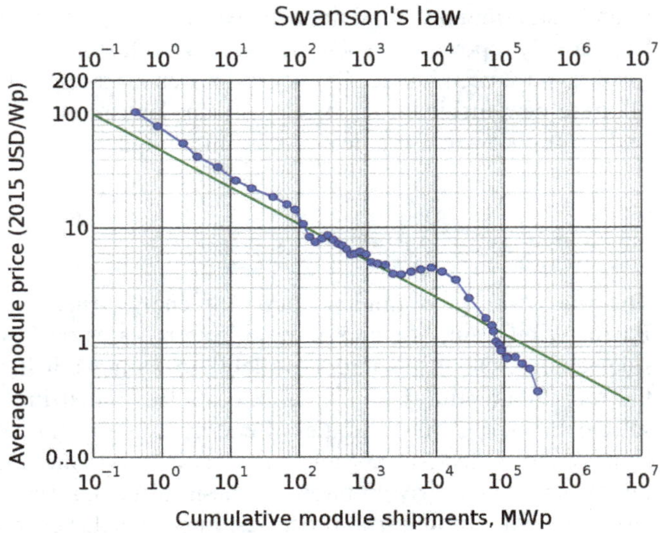

Fig. 5.3 Swanson's effect

For now, the availability of solar energy is only during the day, so there are limited opportunities to power a solar civilization without extensive energy storage. Isaac Asimov once posited the idea of using satellites to capture electricity from the sun and beam it back to a station on the ground, an idea that continues today because arrays can be placed in orbits that do not have a sunset (Fig. 5.3).

The PV industry has experienced tremendous growth over the past decade. Global production was nearly 40 Gigawatts (GW) in 2013, and the cumulative amount of PV modules ever sold passed 140 GW in early 2014, and is approaching 600 GW in 2020. US PV installations in 2014 were 6.2 GW. The figure from the European Photovoltaics Industry Association depicts the tremendous growth in PV installations by regions (Fig. 5.4).

The PV industry measures the volume of modules shipped by the total nameplate capacity of modules sold. The nameplate capacity is the peak power output as measured in direct current (DC) electrical current (i.e., before an inverter converts DC electricity to alternating current—AC). While there is no standard for power output, a typical PV module is generally capable of delivering 200–400 watts. Power output is also used to describe the capacity of a manufacturing facility. A 1-GW manufacturing facility can produce 1 GW worth of PV modules in a year. It is also used to describe the size of a power plant (e.g., 1 GW, 25 MW, 100 kW, etc.) or residential rooftop system (e.g., 4 kW).

PV modules absorb photons, packets of light energy emitted from fusion reactions in the sun, and produce electricity. PV modules generate electric current by the photovoltaic effect, where incoming photons push electrons across a voltage. While there are numerous other technologies that are colloquially called solar power technologies, all non-PV solar technologies utilize the thermal energy from the sun to heat liquids to produce hot water or to generate steam. PV modules utilize photons to generate direct current and do not require moving parts or steam like these other solar powered technologies.

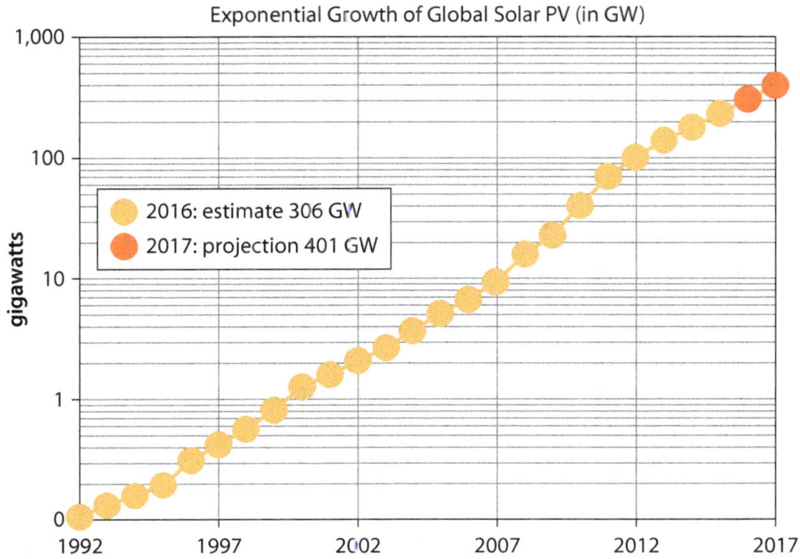

Fig. 5.4 Cumulative PV installations from 1992 through 2018

The US National Renewable Energy Laboratory uses the following definition:

Definition

A photovoltaic module is a device comprised of several photovoltaic cells that is the principal unit of a photovoltaic array. A photovoltaic module's size is roughly 1 square meter.

Nineteenth-century scientists were first to learn of a photoelectric effect. The earliest observance is attributed to Edmund Becquerel who learned about some electro-chemical solutions of copper ejecting electrons when exposed to light. The principles of the photoelectric effect inform the functioning of PV modules, which utilize the PV effect, a term used to describe electrons drawn across a voltage (e.g., when the ejected electrons are carried off as electric current).

Solar energy can be harnessed for heat, motion, light, or electricity generation. Since sedentary agriculture began, and possibly before, heat from the sun was used to dry crops (Smil 2004). In a legend from antiquities, Archimedes was said to have proposed to weaponize solar energy, using a giant lens to produce a ray of heat that could burn the sails of ships of invading navies. Edmund Willoughby Smith further explored this photoelectric effect when it was discovered in selenium bars used to test the electrical resistance of first transatlantic cables in the 1860s (Perlin 1999). Smith learned that the conductivity of selenium was enhanced when the metal was exposed to sunlight, and it became the semiconductor of choice for the first PV cell made of selenium, copper, and gold by Charles Fritts in the 1880s (Perlin 1999). Albert Einstein won a Nobel Prize for his description of the mechanisms—packets of energy—underlying the photoelectric effect, ushering in the quantum revolution.

The PV effect describes electrons drawn across a voltage (e.g., when the ejected electrons are "carried off" or flow as electric current). The PV effect is caused when a semiconductor metal containing extra electrons (an n-type semiconductor) absorbs an incoming photon. This incoming photon raises the energy of the electrons in a metal, elevating the electron energy from the valance band to the conduction band. With electrons freely available in the conduction band, a circuit draws the free electrons toward an electricity load or across the voltage. Conductive metal contacts collect and carry electric current through the PV module to a junction box interconnected to an electrical system. This same architecture with a positive and negative terminal has been used since Fritts' late nineteenth-century selenium solar cell. After Russell Ohl learned the promise of silicon in the 1940s, Daryl Chapin, Calvin Fuller, and Gerald Pearson developed crystalline silicon PVs at Bell Laboratories in the 1950s. Other famous scientists and pioneers include William Cherry, Sheila Bailey, and Sarah Kurtz, the latter two of whom made important contributions to multi-junction solar cell development.

The PV effect is caused when a semiconductor layer with extra electrons absorbs an incoming photon. This incoming photon raises the energy of the electrons in the atoms absorbing the photon and elevates the electron energy from the valance band to the conduction band. With electrons freely available in the conduction band, a circuit draws the free electrons toward an electricity load or across the voltage. Conductive metal contacts harvest and carry electric current through the PV module to a junction box interconnected to an electrical system.

The basic architecture of PV cells is a design called the pn-junction—"pn" refers to positive-negative—the interface between two oppositely charged semiconductor materials. The pn-junction is where a semiconductor with a positive charge is interfaced with one charged negatively. The p-type semiconductor layer conventionally represents the positive side of a solar cell and is depleted of electrons. It has electron-holes where electrons should be. The n-type semiconductor layer has extra electrons.

The pn-junction is needed to "push" the change imbalance and allow electrons to flow from one place to another. When the circuit is complete, this design (presence of electron-hole pairs) will draw electrons out of the n-type and toward the p-type layer. At the same time, holes flow toward spaces formerly occupied by extra electrons. The positive charge in the semiconductor material can be an intrinsic property of the material or can be achieved by chemically "doping" the semiconductor with a material missing electrons. This charge can also be an intrinsic property of the material or extrinsically doped (Fig. 5.5).

Two major categories of PV technologies are crystalline silicon and thin films. Crystalline silicon PV modules are the most common and widespread solar cells, with greater than 90% of the cumulative global market share. There are two types of silicon PV cells. Mono-crystalline silicon semiconductor feedstocks are cooled into one single crystal, while multi-crystalline silicon is cast into crucibles and is

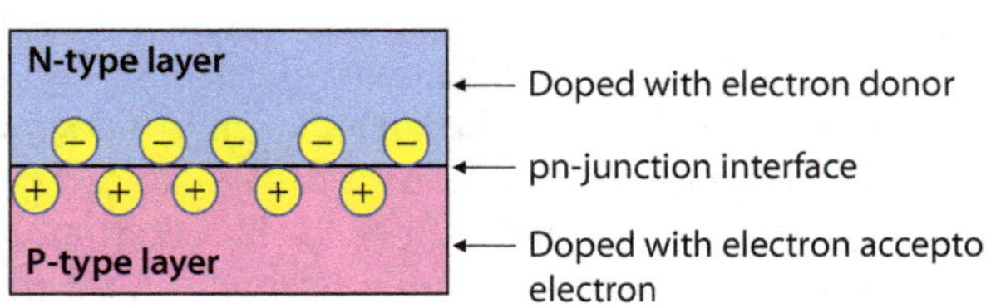

Fig. 5.5 The pn-junction is the interface between the n-type semiconductor and the p-type semiconductor

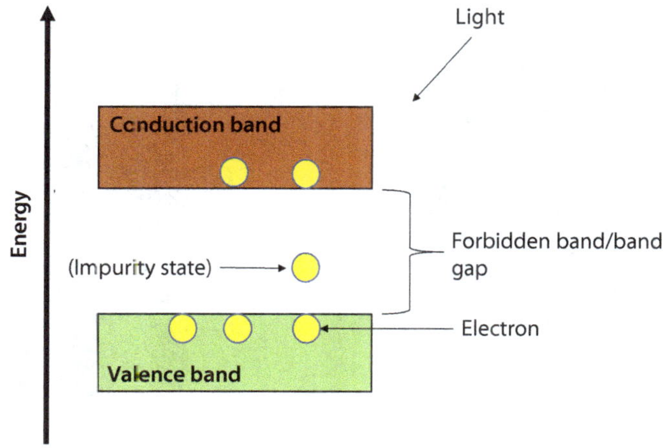

Fig. 5.6 The photovoltaic effect raises the energy level of the electrons when exposed to light. Note there is a an energy level where the electron is forbidden

cooled into a composite of many crystals. Mono-crystalline silicon PV modules are typically more efficient and are premium models in the commercial and homeowner markets. Multi-crystalline silicon PV modules can be 10–20% less efficient, but they have lower manufacturing costs, which is why they used to comprise 70% of the market share of silicon-based PV, but more recently mono-crystalline silicon comprises two-thirds of the supply (Fraunhofer 2020) (Fig. 5.6).

Thin-film PV modules are made from semiconductor materials applied to a substrate like glass or plastic. Satellite images from NASA reveal the immense size of the two largest solar farms in the US. They both use thin-film CdTe photovoltaics manufactured by First Solar, and each produce 550 GW of power. The Topaz Solar Farm is in San Luis Obispo County, California, while Desert Sunlight is in Desert Center, California. They have the advantages of fewer materials, lower manufacturing costs, less manufacturing energy costs, can be recycled, and have faster residence times on the assembly line.

The most common thin-film technologies are cadmium telluride (CdTe), copper indium gallium diselenide (CIGS), and amorphous silicon (a-Si). In 2020, thin films were still only 5% of the PV market share, a decline from previous years (Fraunhofer 2020). CIGS modules have higher theoretical efficiencies than CdTe, but in practice, CdTe has a higher record efficiency than CIGS. Modules and cells of amorphous silicon have been commercially available since the 1970s. They are manufactured by depositing silane, a gas obtained from metallurgical silicon, onto a substrate. Some manufacturers apply an amorphous silicon layer on top of crystalline silicon PV, and others have integrated germanium into the amorphous silicon alloy to improve efficiencies. A few manufacturers have introduced micro-crystalline modules, which are very thin films of crystalline silicon (Fig. 5.7). Perovskites are the darlings of innovation in the solar space in 2020.

Some PV technologies are sold in specialty markets for telecommunications, military applications, and satellites. These are best in remote areas without electricity infrastructure or where generators are not competitive or require too much maintenance. Multi-junction PV modules are stacks of several solar cells (stacks of several pn-junctions) designed to absorb a broader range of the solar spectrum. Some multi-junction PV modules utilize a large glass or plastic Fresnel lens to focus a wide area onto a much smaller solar cell surface. The most widespread

Fig. 5.7 Desert Sunlight Solar Farm, Desert Center, and Topaz Solar Farm in California Valley, both are in California

semiconductor materials for multi-junction solar cells are gallium arsenide, indium gallium phosphide, germanium, and indium phosphide. Dye-sensitized, perovskites, and organic PV cells remain limited to laboratory research and at pilot-scale manufacturing. One hybrid solar technology, called concentrated PV, concentrates light using a lens onto a relatively small solar cell. Science fiction writer Isaac Asimov's (1941) idea of placing solar arrays into space and sending energy back to Earth via radio waves was proposed in a project that received a power purchase agreement from California investor–owned utility Pacific Gas and Electric Company in 2008, but the project was never developed. The space-based projects would be appropriate for multi-junctions solar cells' additional cost.

The semiconductor supply chain for crystalline silicon PV starts with metallurgical grade quartz. Quartz is mined and coked into metallurgical grade silicon, much of which is sent to steel foundries. Metallurgical grade silicon is used to manufacture polysilicon, the input that is purified and shaped into silicon wafers. Polysilicon is made from processing metallurgical grade silicon into chlorosilanes. Polysilicon plants resemble refineries to the casual observer and is typically produced in places zoned for industrial chemical facilities, whereas other downstream phases of manufacturing and assembling can occur in light industrial or commercial zones. The chlorosilanes are processed in a reactor where pure polysilicon is deposited on crystalline rods. The rods are eventually removed and broken into polysilicon chunk—the silicon feedstock sent for preparation into silicon wafers.

Making crystalline silicon wafers next involves melting polysilicon chunk. Mono-crystalline silicon is formed by the Czochralski method where the higher temperatures and slower cooling rates result in extremely pure crystalline silicon ingots. Multi- or poly-crystalline silicon is made by melting polysilicon in crucibles, where the cooling method results in multi-directional crystals in the silicon ingots. The ingots are usually doped to give them an intrinsic charge (this can either be p-type or n-type). Beyond the steps where polysilicon is melted into either mono- or multi-crystalline silicon, all steps are the same for PV manufacturing.

The ingots are next sliced into bricks of silicon and then into individual wafers, using diamond wire saws submerged in a slurry mixture. The wafers are etched and cleaned to remove sawing damage. Since the ingot preparation stage usually adds one charge, cells typically require chemical doping on one side of the cell surface so each cell has one p-type and one n-type side. This charge is achieved by diffusing chemicals into the top several millimeters of the PV cell. An anti-reflective coating is added next with a mixture of ammonia and silicon nitride to maximize the absorption of incoming photons. Finally, the cells are fused together with metal contacts so that the cells are able to collect electrons for electricity. The cells are finally sorted and checked for quality control.

PV cells are finally assembled into a PV module. In this stage, the solar cells are laid out on a backsheet and interconnected. Solar-grade glass, the bulk of the module's mass, is added on top of the cells and backsheet. The glass is finally covered and protected with an encapsulant. A junction box is added to the box, allowing for electricity interconnections to be made. If the PV module design calls for a frame, this is added last.

Thin-film PV modules are an alternative module design based on thin films of semiconductor materials applied to a substrate like glass or plastic. They have the advantages of fewer materials, lower manufacturing costs, less manufacturing energy costs, and faster manufacturing times. It is still widely speculated that thin films still have theoretically lower cost potential.

PV module, many thin-film manufacturers have still not been able to compete with crystalline silicon on a price-basis with the exception of First Solar. The most common thin-film PV technologies are cadmium telluride (CdTe), copper indium gallium diselenide (CIGS), and amorphous silicon (a-Si).

How Much Energy Does Earth Get from the Sun

Earth is bathed in electromagnetic radiation from the sun. How much can be harnessed for electricity generation? To understand this, we need to know a few things about the sun and Earth, and the energy that travels across the solar system. First note that solar flares, cosmic rays, and microwave radiation also interact with the Earth. But the amount of energy they deliver to the planet is insignificant compared to the heat and energy of light from the sun.

To understand the solar resource, start with the major flows of energy from the sun. Overall, 31% of the energy that hits Earth is reflected. A total of 45% goes directly to heating the atmosphere. Another 23% goes to evaporating water, while the remaining 2–3% is absorbed to become the energy of winds or photosynthesis.

How much of the energy that hits Earth can be harvested as electricity? Start with the Earth's circumference, and divide that by 4 = 10^7 meters. From geometry, calculate the radius and think of the solar energy delivered to the planet as interacting with the surface area of a circle. There are 365 ¼ days in a year considering the leap year.

$$\begin{aligned} &1.733\times10^{17}\,\text{J/second}\\ &\times 86,400\ \text{seconds/day}\times 365.25\,\text{days/year}\\ &=5.46\times10^{24}\,\text{J}=5,460,000\,\text{EJ/year} \end{aligned}$$

The global rooftop area of the world is 3.8 × 10^{11} m^2. What portion of Earth is this and how much global energy generation per year could be accomplished with the fleet of global rooftops?

Compared to other sources of energy, this is a far greater resource than any other reported earlier (Fig. 5.8).

Net metering is a simple model where prosumers generate their own electricity, and only pay for what they import from the grid. In times of excess, they can also export energy back to the grid for credit. Since the Energy Policy Act of 2005, every public electrical utility is required to offer net metering. Net metering has been criticized because generators do not pay for fees associated with transmission and distribution, and this is thought to shift those costs to non-generators (Fig. 5.9).

5

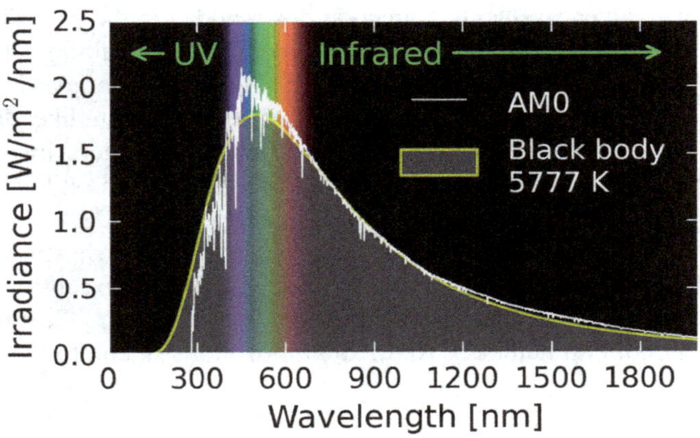

Fig. 5.8 The energy spectrum from sun to Earth and the bandwidths available for solar power

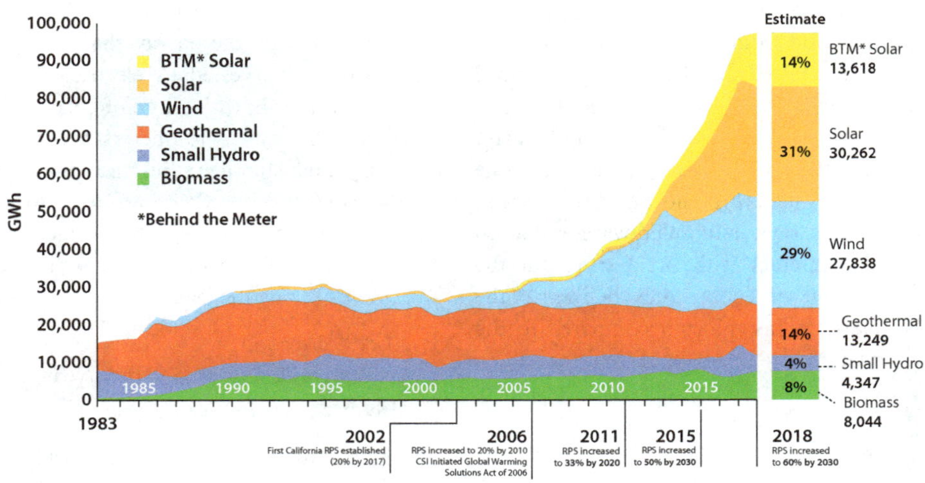

Source: California Energy Commission, staff analysis November 2018

Fig. 5.9 California renewable energy generation, 2010–2017

Energy storage can present a good solution to the challenges facing future electricity grids. Demand response tools and smart appliances will utilize energy off-peak, such as an air-conditioning system pre-cooling a home or a hot-water heater turned off a few minutes, only losing a degree or so. Virtual net metering is a model allowing off-site renewable energy generation credits to be used for homeowners (▫ Figs. 5.10 and 5.11).

Land-use change from utility-scale solar energy (USSE) facilities can impact wildlife and lead to habitat loss and/or fragmentation (Lovich and Ennen 2011). The specific location of USSE projects determines the severity of ecosystem impacts. California's desert land types most commonly disturbed for USSE facilities are shrublands and scrublands (26%), and less than 15% of such facilities are sited on "compatible" lands (Hernandez et al. 2015). But this region is a biodiversity hot spot. The impacts in other socio-ecological systems could be different (▫ Fig. 5.12).

Solar hot-water heating offers a different application of solar power to displace higher carbon fuels. In some places, distillate fuel oil is used for water heating; in others, it is electricity, natural gas, or propane (▫ Fig. 5.13).

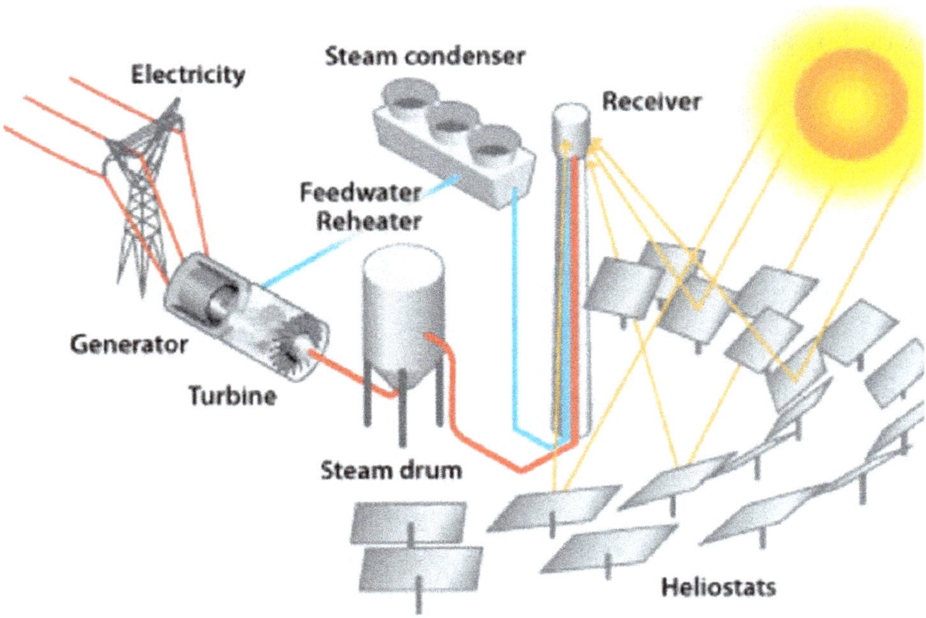

▫ **Fig. 5.10** A concentrated solar power tower uses heliostats to produce steam

Fig. 5.11 Solar power tower in Ivanpah Valley, California. (Photo: Mulvaney)

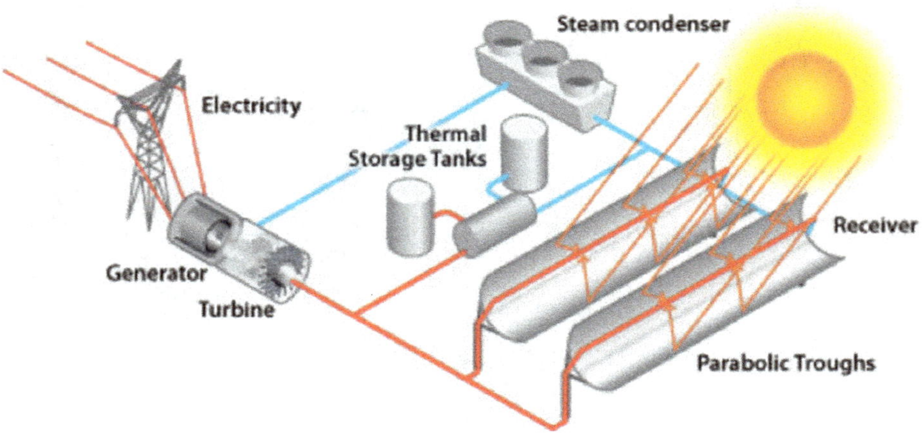

Fig. 5.12 CSP parabolic troughs produce steam by using large curved mirrors

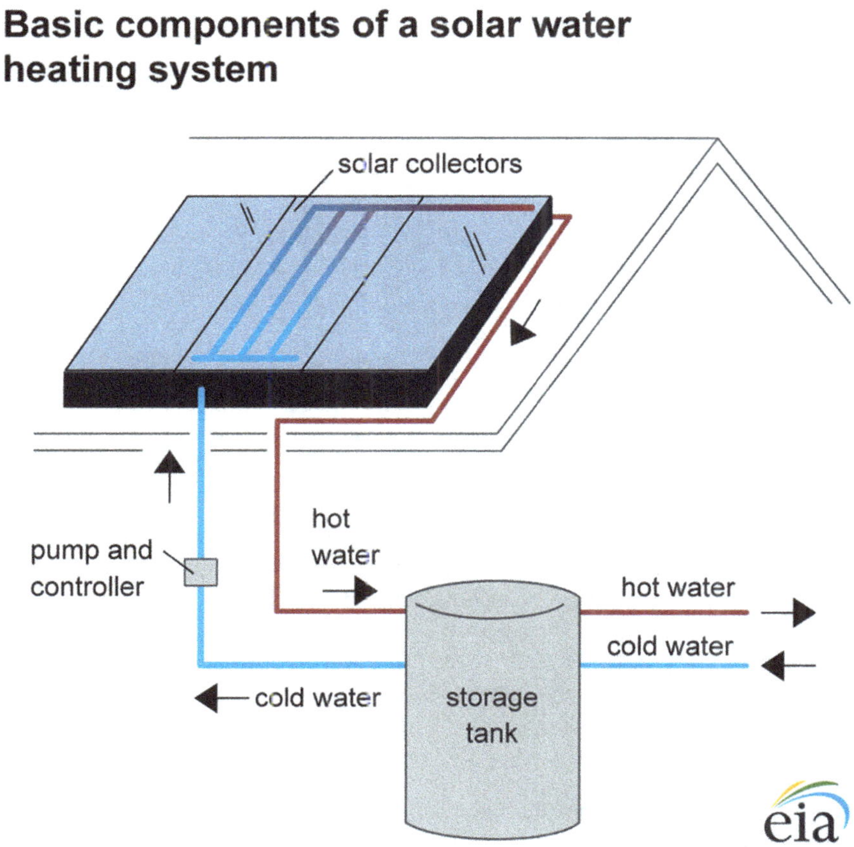

Fig. 5.13 Flat plate solar hot-water heater. Note: This is a simplified diagram of a drainback-type solar water heating system. (Source: U.S. Energy Information Administration)

Assignment 5.2 Solar Hot Water Sizing and Cost

In this exercise, you will be introduced to the National Renewable Energy Laboratory's System Advisor Model (SAM). This is a free tool and arguably the most comprehensive solar assessment tool available, which can be used for assessing concentrating solar, PV, solar hot water, and more.

1. (a) Estimate the *daily thermal output* required for a 3-person home that uses 60 gallons of hot water per day. The water comes into the home at 51 °F and is warmed to 120 °F.
   (b) On a cloudy day, the system will only produce 12,000 Btu. What percentage of the hot water can be heated with solar energy on this cloudy day? Multiply the difference in temperature and volume by 8.33 Btu /gallon °F.
   (c) If hot-water use is cut by 20% and the hot-water temperature

goal turned down to 117 °F, what would be the daily solar thermal output required?

2. Find the maximum collector temperature for an 8 m^2 solar hot-water heater in the following scenarios: (a) sunny summer day at 32 °C with solar insolation at 1,000 W/m^2 and (b) a cloudy summer day at 25 °C with solar insolation at 300 W/m^2. The loss rate is 8.5 W/m^2 per °C and the panel lets in 92% of incident sunlight. On the sunny day, (c) estimate the power input (kW) that goes into the panel at peak insolation (1,000 W/m^2), (d) the power lost from the surface, and (e) the net power gain overall.

5

Assignment 5.3 System Advisor Model (SAM) Part 1

3. Download the System Advisor Model (SAM) from ► https://sam.nrel.gov/ and use the model to estimate the cost of buying a solar hot-water system and displacing electricity. Start SAM and select "solar water heating" and "residential – distributed." Click OK.

Location and Resource

Go step by step, filling out the known information. For location and resource, search for CA San Jose Intl Ap and select the weather/insolation data.

Solar Water Heating

Estimate your average daily hot-water requirements in the "solar water heating" section. Leaving the default is fine. How much did you enter? _______kg (Note that you have to convert gallons to kg of water).

Under Collector, leave all of the defaults. This is an average collector. Note that SAM stores a library of solar thermal hot-water collector models.

In the Solar Tank and Heat Exchanger section, set the solar tank maximum water temperature to 105 °F and the outlet temperature to 60 °F.

System Costs

Let us presume that in the future, the costs will be cheaper due to economies of scale, manufacturing efficiencies, and material improvements. Set the collect costs to be **200 $/m²** and storage to **750 $/m³**. Some policymakers feel that solar water systems should not have a sales tax. Go to Indirect Capital Costs and change the percentage of the sales tax to 0% of direct costs.

Degradation

Nothing needs to be changed for degradation.

Financial Parameters

We are going to assume this is paid for with cash. To do this, go to loan parameters and set the debt fraction to 0.

Incentives

Under incentives, go to Production-Based Incentive (PBI) and set the Amount ($/kWh) next to Utility to **0.54** and the years to **10**.

Electricity Rates

Search for your home zip code and find your local utility rate. "download and apply utility rate." Once they apply, uncheck the box for net metering.

Electric Load

Leave as is. Click **"Simulate"**

- Annual energy saved _____
- Solar fraction _____
- Pay-back period _____

If you want to know the costs of a system that uses natural gas instead of electricity, click Macros (below Simulate) and Auxiliary gas heater. Set burning efficiency at 90% and load the tank loss factor the same. Run the macro-portion.

5.4 Hydropower

Hydropower was first used in China over 2,000 years ago to crush grains, break ore, and other simple mechanical processes. In the nineteenth century, hydropower would be used to generate electricity. From there, the use of hydropower quickly spread across nations, with the US and Canada leading the way. Today, the world's hydropower production is dominated by China, Canada, and Brazil, with China producing 1,064 TWh, nearly three times as much energy than the next highest-producing country (IEA 2018).

Hydropower is power that comes from the falling or flow of water. Most hydroelectric power plants work by transferring potential energy to kinetic energy to electricity or simply by just transferring kinetic energy to electricity. Hydroelectric energy comes from capturing the power of moving water; as the water is released from the reservoir, it flows through a turbine, making it spin. The spinning generates kinetic energy, which is converted to electricity; the motion activates the generator, by producing electricity for the power grid.

The amount of power generated by a specific dam and river system depends on several factors. The key variables are the flow rate, the height (sometimes referred to as the head), and the efficiency of the system. The efficiency of the system can be reduced to four dimensionless factors—the efficiency of the turbines, the flow through the penstock, the gears, and the turbine. The efficiency factors are multiplied together to get an overall flow rate.

Hydropower is renewable insofar as the flow of water is dependent on the sun and the water cycle. Hydropower is also versatile, as dams can control the flow of water and electricity generation. Dams can go from producing no electricity to giving maximum output in an instant, making energy available on demand. Farmers can also use the stored water to irrigate crops. The reservoirs of hydroelectric plants can be locations for recreational activities such as boating and fishing, potentially stimulating the economy in the area. While hydropower has many benefits, it also has many environmental costs.

A total of 84,000 dams were built in the US in the twentieth century (Reisner 1993). Legendary North American folk musician Woodie Guthrie was hired by the Bonneville Power Administration in the 1930s to travel and write about the people, communities, and landscapes of the Columbia River Valley. Many of the dams that were built before the 1960s are without hydroelectric power, an estimated 80,000 unpowered dams. Research often finds trade-offs between ecological values and energy. One study found that dams in streams cause less harm to migratory fish, but dams in larger waterways have a greater energy value because of the larger flow and storage capacity (Jager et al. 2015). These results informed a few design principles for dams aimed to increase ecological sustainability, including adding freshwater reserves around the remaining watershed catchments, building dams in areas of tributary watersheds and not in the main downstream areas of rivers, making sure that habitats are provided between dams that will encourage and support species reproduction and preserve offspring, and finally, making decisions at the scale of river basins. Some of the benefits attributed to large hydropower dams are displaced fossil fuels, storage for urban water

supplies, flood control, irrigation, and even the creation of jobs. Hydroelectric power generation has some major impacts, including loss of habitats, forests, streams, restriction of river flow, and negative effects on aquatic species like salmon (Abbasi and Abbasi 2000).

Hydropower can change stream flows, increase turbidity, displace or lock-in habitats, change fish migration patterns, degrade wilderness resources, and increase erosion and landslides (California Energy Commission 2020a, b). Not only do large dams cause ecological, environmental, and social impacts, but they are also a large financial burden with costs that have a habit of going over budget during construction and for mitigation efforts that can drag on for years. However, dams can be relatively inexpensive sources of low-carbon power.

An important area undergoing energy transitions toward increased hydropower is in eastern India, in the Himalayas (Rampini 2016). In Brazil, there is the River Xingu basin in Amazonia, where dams are promoted as clean energy sources while they transform ecosystems and livelihoods (Hall and Branford 2012). Some researchers have identified different development practices and have suggestions for indigenous consultation, consent, and incorporation, if not leadership, in planning (Kelly 2019).

In the US, the Federal Energy Regulatory Commission (FERC) is the authority that determines and issues licenses for the operation of non-federal hydroelectric power facilities. FERC was not required to include environmental protection into its decision-making for project licensing until the passage of the National Environmental Policy Act in 1970, the Clean Water Act in 1973, and the Endangered Species Act in 1973 (Schramm et al. 2016). FERC was required to consider environmental impacts and "protection and enhancement of, and mitigation of damage to, wildlife, environmental quality, and recreational opportunity" after the passage of the Electric Consumers Protection Act of 1986 (Schramm et al. 2016). Even after the passage of these laws, FERC was further required to include the US Fish and Wildlife Service or National Marine Fisheries Service's fishway prescriptions. They were also required to set a minimum for stream flow that was included as part of state water quality certification (Schramm et al. 2016). All of these changes have greatly increased licensing time, and they have also made it more difficult to meet mitigation requirements.

Dams create water reservoirs that function as water storage, flood control, and recreational use. Run-of-river facilities divert water from natural channels of water to a turbine and then return the water to the stream or river after it has passed through the turbine. Pump storage facilities pump water at off-peak demand hours from a reservoir at a lower elevation to a reservoir at a higher elevation for storage; the water is then released during peak demand hours to turn the turbines for power generation.

Dam failures throughout history have been the site of many human disasters. In Banqiao, China, a dam failure aided by copious amounts of rain from Typhoon Nina in 1975 killed over 200,000 people and left 11 million homeless. Even ancient civilizations faced these types of tragedies. The Great Dam of Marib in present-day Yemen successfully held back the waters from about 800 BC to about 600 AD when the dam was breached. In, 1889, a disaster in Pennsylvania killed 2,209 people, when the great Johnstown Flood took down the reservoir and dam system. In 1928, the St. Francis Dam, near Los Angeles, killed over 450 people when it collapsed and swept through downstream communities and many migrant camps in the middle of the night.

Dams are low-carbon energy sources, but not zero carbon. Reservoirs may produce considerable amounts of GHGs in the form of CH_4 because of the plant material decaying in flooded regions, known as anaerobic environments. Reservoirs also collect various types of sediments that concentrate nutrients and pollutants, resulting in the death of wildlife and compacting waterways.

There are new technologies available that can provide power by installing in-pipe turbines near water facilities. There is technology available that produces renewable energy, while at the same time not having a permanent impact on the site of installation. Spherical turbines can be installed inside gravity-fed water pipelines. A system like this is installed for the Portland Water Bureau in Portland, Oregon. Using four turbines, the micro-hydro system produces 900 MWh per year for the electricity grid (Fig. 5.14).

Fig. 5.14 Bonneville Dam on the Columbia River. (Photo: Mulvaney)

What are the opportunities for sustainable hydroelectric power production at smaller scales? Can hydropower be used as a domestic renewable energy to run appliances, for heating, or cooling a home? These developments are usually described as pico-hydropower or micro-hydropower. The sources can be easily obtained as one of the simplest and most consistent forms of renewable energy, depending on the local climate and competing water demands. Micro-hydropower is best where water flowing across a landscape is present. These systems can produce between 5 kW and 100 kW, depending on the resource and design. This means conversion of river flow into electricity, using gravitational potential energy. A micro-hydropower system can easily generate power for smaller farm, resorts, or large homes.

With pico-hydropower, the turbines are even smaller and produce around 10 kW to similarly generate enough energy to power the same living standards as does a micro-hydropower system. Pico-hydropower systems are rarely fed into the power grid, but instead use electricity on-site or store it. Attaining the water source for pico-hydropower usually begins with identifying where water enters and exits from a creek, whether any permit is required or there are water rights issues, and where to locate how the water enters and exits the household. Other potential sites of pico-power include rain gutters, sprinklers, hoses, dishwashers, sinks, showers, toilets, and washing machines.

One of the major changes facing hydropower in the face of climate change is snowpack and the availability of water. Some climate models project snow levels rising to higher levels in some parts of the world. This could result in less available water during the driest parts of the year and directly translate to less hydropower available.

5.5 Bioenergy: Biofuels, Biomass, Biogas, and Biochar

Biological processes are a potential energy source for decarbonization. Plants, animals, fungi, and other species on Earth fix and store energy from the sun and carbon in the atmosphere through photosynthesis. On Earth, photosynthesis stores energy at a rate of 133 TW, which is about ten times global power consumption. At any given time, 1.5 zetajoules (ZJ) energy is stored in biomass or approximately 35 times annual energy consumption. If it is not used as an energy source, most of the energy stored in biomass will be eventually radiated as low-temperature heat, and carbon will be released from the organic materials.

Definition

Biofuels are any organic matter available on a renewable or recurring basis. This includes agricultural crops and trees; wood and wood wastes and residues; plants (including aquatic plants); grasses; residues; fibers; and animal wastes, municipal wastes, and other waste materials.

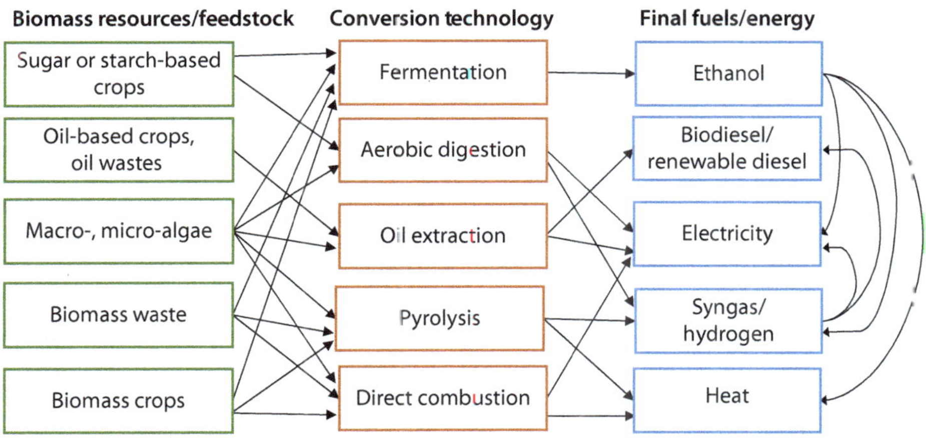

Fig. 5.15 Different platforms for biofuels to energy

Photosynthesis occurs in the presence of light with the following reaction (Fig. 5.15).

$$6\,CO_2 + 6\,H_2O + \text{light} \rightarrow C_6H_{12}O_6 + 6\,O_2$$

Biofuels are solid, liquid, or gaseous fuels produced from organic, carbon-based, materials. Liquid biofuels include biodiesel and ethanol, and some higher-order biofuels such as biobutanol. Energy crops are high-yield production crop species. These are higher energy content but more expensive to produce because they require more inputs. They can roughly be divided into sugar crops, grains, and oil seed crops. Biowastes refer to fuels that are the waste products from other activities. These are typically lower in energy content but plentiful in supply. These include agricultural residues and other plants, ranging from trees to seaweed to grasses.

Ethanol is the most common liquid bioenergy source and is sometimes called the sugar platform for biofuels. It is produced by biochemical conversion of fermentable sugars or starches that can be converted to fermentable sugars with other biocatalysts.

Ethanol fermentation from sugars reacts as follows.

$$C_6H_{12}O_6 \rightarrow 2\,C_2H_5OH + 2\,CO_2 + \text{heat}$$

Blending ethanol into gasoline to make it burn cleaner and with fewer carbon emissions has raised questions around land use, carbon balance, water use, and water quality impacts. The fuel mix already comprises about 10% of light vehicle energy use in the US where it mainly comes from corn, and also wheat. Ethanol plays a larger role in Brazil where it is made from sugarcane in farms in the eastern-most seaboard of the country near the megacity Sao Paulo. Numerous alternative feedstocks such as micro- and macro-algae, waste-to-ethanol, show better environmental performance than corn and sugarcane. Ideas to power transportation by ethanol dates back over 100 years when a vision advanced by Henry Ford argued that farmers should grow the fuel for automobiles.

Most experts believe that increased ethanol production and environmental performance will require advances in more sustainable and cheaper second- and third-generation biofuels. Cellulosic biofuels—known as the plant products platform—will offer the opportunity to turn any plant materials into fuels.

Example 5.1 How Much Land Is Needed for Bioenergy Powered Civilization?

Use the following parameters to approximate the amount of land needed to produce all the energy needed to power all human activities. The conversion of solar energy into liquid biofuels is inefficient compared to directly making electricity from photovoltaics. Biofuels require large areas because only some percentage of incoming energy flow is converted into liquid fuels.

- 3 TW requires $\approx$ 600 million hectares = 6×10^{12} m^2
- 20 TW requires $\approx 4 \times 10^{13}$ m^2
- Land with biofuel crop production potential is about 2.45×10^{13} m^2
- Total land area of Earth: 1.3×10^{14} m^2
- Hence, it requires 4/13 = 31% of total land area of the Earth to produce enough biofuels to meet society's energy demand.

The land-use intensity of biofuels and possibility of profiting from their development is leading to land grabbing. Land grabbing is when organizations might use their "green" credentials to appropriate or alienate local farmers from their land (Fairhead et al. 2012). Lands have been sold all over the world to large companies that operate agricultural production systems. The result can lead to further marginalization of the dispossessed (Li 2018).

There are a number of categories of bioenergy that are referred to as biofuels, biogas, biochar, and so on. Wood was the primary energy source for human civilization, dating back before the fossil fuel era. But even today wood is widely used across both the developed world, in places that lack basic energy infrastructure, or where wood stoves are used in residential heating. In the developing world, these wood dependencies are for all sources of energy, including cooking and all indoor heating, whereas in the developed world heating from wood can be both a winter necessity or common heating fuel for a region, or, in some places, it may be a rarely, but seasonally, used source of heat.

Biomass for electricity in some places can be a meaningful contributor to renewable energy supplies. The most common feedstocks for combustion include wood pulp, crop residues, forest residues, and organic waste. Biomass is likely to remain a niche energy source as the land and water requirements to make meaningful contributions are very high. Today, there is a thriving biomass for heat economy, where wood grown on plantations in Southeast US is shipped to Northern Europe.

Of the climate solutions to the use of wood for energy, biochar production is often described the best way to avoid emissions of carbon into the atmosphere, that it has unclear economics and many uncertainties. Biochar production for soil carbon sequestration could lend to permanent and verifiable offsets, as well as relatively straightforward carbon accounting methodologies. Carbon sequestered in biochar is more durable than biomass carbon sequestration because carbon remains locked in soil for a duration that is orders of magnitude greater; estimates vary from 2,000 to 50,000-year half-life, depending on temperature of preparation and method of incorporation into the carbon saving media.

Breaking down biochar is not dependent on fluctuations in microbial decomposition rates due to climate change, unlike soil organic matter. Other forms of carbon sequestration have higher uncertainties around future carbon storage than biochar due to the complexities associated with the fate of forestlands and potential surface leakage from geological storage, in which storage mechanisms are not well understood. These uncertainties have made many forms of carbon sequestration controversial because of problems with quantification, verifiability, additionality, and permanence.

There are important co-benefits of biochar, including increased soil fertility, water retention, and renewable energy production as a by-product of biochar manufacture. Despite the high potential of biochar to sequester carbon, a significant amount of information will be necessary to design a comprehensive and justifiable offset protocol—sometimes counting nature can be fickle. Multiple factors, such as feedstock type, pyrolysis method and temperature, and biochar form, proportion, and size will affect the derived benefits and the longevity in soil.

The key to understanding the carbon balances for feedstock production requires a review of carbon fluxes in natural and agroecosystems, land-use change dynamics, and the optimization of yields. Potential biochar feedstocks include (1) forest waste or slash from timber management or from fuel reduction strategies; (2) agricultural residues such as olive pits, corn stover, nutshells, rice husks; and (3) fast-growing woody crops like poplar, willow, eucalyptus, as well as some grasses. The trick here is that some of these sustainable crop characteristics also make them weedy and possibly invasive, which maybe means they are not so sustainable if not properly managed from an ecological perspective.

The question of additionality as it relates to claims about carbon reductions asks whether a project would have happened in the absence of the monetary value placed on sequestration. Since biochar value is only gained through incentives for carbon sequestration or energy production, it is unlikely that biochar projects, particularly those that optimize biochar production at the expense of energy production, would have happened without some offset program in place. There are other economic considerations that may contribute to the economic potential of biochar production, such as waste disposal costs. A balanced approach would evaluate the costs of projects in the presence and absence of available offsets for carbon sequestration, including an analysis of the trade-offs between energy and biochar optimization.

Biogas is generated from the anaerobic decomposition of organic matter. Biogas refers mainly to biomethane (also CH_4), which are usually formed by biological organisms called methogens that thrive in anaerobic (absent oxygen) environments. But biogas could refer to several other biologically derived gases, including hydrogen and bio-syngas. For the purposes of this chapter, biogas will refer biomethane exclusively. Biogas is often used in energy production or heating. Biogas is composed of mostly CH_4 and carbon dioxide (CO_2), and small amounts of nitrogen, hydrogen, and carbon monoxide.

An important method of CH_4 generation is called the Sabatier reaction. While it is widely known, it is still being studied and has not yet been implemented in a utility-scale production setting; however, some feel that it holds much promise. In the Sabatier reaction, CO_2 reacts with hydrogen to form CH_4 and water.

With the advancement in technology, such as the adoption of anaerobic digestion, the production and collection of biogas is now easier than before. The use of biogas has been growing rapidly, especially in Europe. The rapid rate may also have contributed to other problems. According to a study conducted in 2016, over 200 accidents related to biogas plants have been reported in Europe, with majority of the events related to fire caused by biogas (Trávníček et al. 2018).

One common source of biogas is from anaerobic digestion of biodegradable organic feedstock such as cow manure or organic matter collected at wastewater treatment facilities. In order to obtain biogas, farmers collect animal manure and put them into airtight containers. The biochemical process takes place when the inside temperature of the container increases to a controlled range, then the bacteria start to digest the organic matter. The manure has four different components—water, digestible organic material, indigestible organic material, and inorganic material. Inside the sealed container, water and digestible organic material from the manure will form biogas after being broken down by bacteria during the digestion process, while the indigestible organic material and inorganic material will become effluent (solids and water). The biogas created in this process can then be used as an energy source to heat water or can be transformed into electricity on-site.

GHGs from agriculture accounts for almost half of the total nitrogen (mostly N_2O) emissions and over a third of the total methane (CH_4) emission. Studies reveal that the 20-year global warming potentials for N_2O and CH_4 are 275 and 62 times higher than that of CO_2 (Pattey et al. 2005). Using animal waste to produce biogas also can displace fossil fuels. Livestock waste from cows has the best potential for producing biogas. Manure from about 1,500 cows can sustain about 200 kW of power.

Dairy pilot projects utilize a technology known as “dairy digesters” to capture and process biomethane gas from manure to produce renewable natural gas and incorporate it into the already existing pipeline network. Dairy digesters are anaerobic manure digesters and take advantage of a biochemical process known as anaerobic digestion to convert the energy stored in manure into CH_4.

Definition

Anaerobic digester—a reactor that with increased pressure and/or temperature can react organic matter to produce CH_4.

Biogas production from dairies will require incentives for owners to produce biogas as digesters can be expensive. It can also complicate local air and water pollution. Biogas emissions include nitrogen, sulfur dioxide, and hydrocarbons. There are about 9 million dairy cows in 75,000 dairy herds in the US with nearly 2 million cows on 1,500 dairy farms in operation in California (USDA 2015). A 1,000-lb dairy cow creates an average of 80 pounds of manure each day. Feed type and temperature affects the amount of biogas that can be generated from manure.

Biogas is also produced at landfills. The carbon source in the methane is sometimes biological and at other times can come from a fossil source, making the definition of this source of renewable energy a little hazy. There are many political debates about whether this technology should be classified as renewable in energy policy because of the pollution caused and its proximity to environmental justice communities as described with incinerators by Geographer Dr. Ingrid Behrsin (2019).

Most municipal solid waste contains biomass such as paper, cardboard, wood, plastics, food, and so on, which decompose creating CH_4, CO_2, and hydrogen sulfide. Capturing and combusting CH_4 before it is released has some climate benefits, as the escaped CH_4 produces the GHG effect at 25 times higher than that of CO_2. Landfill biogas recovery can yield substantial energy, economic, and some contested environmental benefits. For every 1 million tons of MSW, approximately 0.8 MW of electricity or 432,000 cubic feet per day of landfill gas (LFG) can be produced. Also emitted by landfills is leachate, which is a by-product from landfills. Leachate is produced from the precipitation falling onto landfill cover, which seeps down to waste, causing it to percolate up through landfill cover. Leachate needs to be treated and disposed of. The facilities that have recovery systems in place capture CH_4 for the use of creating energy through turbines, microturbines, or maybe, someday, using Stirling engines. Of all the landfill gas produced, 60% or more is not captured because of the lack of collection systems, and those that do continuously lose an average of 25% (Randolph and Masters 2008). Avoiding landfilling biomass can avoid associated GHG emissions. Unfortunately, closed landfills continue to produce landfill gas for roughly 100 years from the previously buried biomass but at a diminishing rate over time (Rajaram et al. 2011).

Biogas is a product from organic waste. Improper management of the organic waste and the digestates can have adverse impact on the environment. Regulations applicable to organic waste treatment and recycling also apply to the processing of biogas and the digested manure. With the increased popularity of biogas as a climate action strategy for manure, it will be important that regulations stay on top of the issue.

Incineration is another process used for converting waste-to-energy of MSW. Incineration facilities receive residual waste from landfills or waste transfer stations. With incinerators, the type and composition of waste determines energy content. If organic waste decreases and there is an increase in plastics and rubber, it results in higher energy values. Waste initiatives, diversions, and consumer purchasing habits and home waste diversion practices (composting) could dictate energy output from biomass. Incinerators equipped with air pollution control systems can reduce air pollutant emissions, but not all of them. The main emissions of incineration are CO_2, water, and ash, but it can also include toxic heavy metals, dioxins, furans, and others chemicals. Ash left over in the combustion chamber is discarded at a hazardous waste landfill. Hauling waste to incinerator facilities and back contributes GHG emissions via transportation.

5.6 Geothermal

Geothermal energy is accessible everywhere on Earth. Earth loses heat at about 40 to 50 TW each year, providing an average of 0.1 watts per square meter. The sources of this heat include friction from fluids moving in Earth's core, radiative decay of elements in the mantle, and pressure from the planet's weight. Geothermal hot spots that make it to Earth's surface include hot springs and the famous "Old Faithful" geyser at Yellowstone National Park.

There are three geothermal power plant technologies used to convert steams coming from underground into electricity—dry steam, flash steam, and binary cycle. One of the most efficient types of geothermal technology used today is the dry steam geothermal power plant. Dry steam plants apply steam directly to a turbine, which drives a generator that generates electricity. The type of geothermal power plant is dictated by the nature of the resource. The geyser complex of 22 geothermal power plants located in the Mayacamas Mountains of Northern California is the world's largest single source of geothermal power for electricity.

Flash steam plants are the most common type of geothermal power generation plant in operation today. Like dry steam plants, they use a fluid that is underground at temperatures greater than 360 °F and pump it under high pressure into a tank. The tank at the surface is held at a much lower pressure, causing some of the liquid to vaporize at a rapid rate. From it, the vapor turns the turbine connected to an electricity generator to generate electricity for usage. Binary-cycle geothermal power generation plants are different from dry steam and flash steam systems technology because the fluid from the geothermal reservoir never comes in contact with the turbine or generator. Binary-cycle power plants use a closed-loop system. Like the two other power plants, the binary-cycle power plants virtually emit nothing to the atmosphere. It is predicted that a significant proportion of geothermal electricity in the future could come from binary-cycle plants because most of the geothermal energy is below 300 °F, a resource level that favors binary cycle.

The potential of global geothermal energy growth is significant. Sites along the Pacific "ring of fire" or other places near active volcanic areas offer significant potential for electricity generation. Geothermal energy provides more than 15% of electricity in several countries, including Iceland, Kenya, El Salvador, Costa Rica, and the Philippines. Iceland's Blue Lagoon is also a popular swimming attraction for tourists. Baths are located directly in the shadows of the power generation facility. One major use of geothermal power in Iceland is the production of aluminum.

While it is widely accessible, the economics for geothermal energy is far more complicated. Geothermal reservoirs can be of different quality that can pose site- and rock-specific challenges to the siting process.

Like most things that require taking material or fluid from Earth's crust, geothermal power plants must manage sulfur emissions. When a volcanic eruption in Hawaii led to lava encroachment at a geothermal facility on the big island in 2019, there was concern about deadly hydrogen sulfide emissions. The good thing is that some are proposing extracting some important metals from geothermal brines, notably lithium, rare Earth elements, and uranium.

One innovation in this space is called enhanced geothermal systems (EGS). With this technology, the heat reservoir rock is fractured and cracked to maximize the surface area of hot rock as water is passed over it. The development of EGS technology could make a much larger resource available, though this EGS process may be linked to increased potential for earthquakes and may be too expensive.

Low-grade geothermal heat just a few feet below ground can also be used more directly to heat buildings and homes. Ten feet down below the Earth, the ground temperature is consistently between 10 °C and 16 °C . This resource can be tapped to provide cooling in the summer and heating in the winter for homes and large buildings. Geothermal heat pumps work by burying a loop of pipes called heat exchangers just below the ground surface. Filled with water or antifreeze solutions, pumps will circulate the heated fluid from inside to outside and the other way around, saving money on air-conditioning during winter and summer.

5.7 Wave Power

There are many different designs for a technology able to extract power from waves. Wave energy can be extracted from surface waves, pressure fluctuations from water surface, or from full wave; some technologies stay in a fixed position and let waves pass them, and some follow the waves and move with them. Waves vary with temperature and pressure differences. Using a wave energy converter (WEC) or some type of hydraulic motor is necessary to harness wave energy and generate electricity.

Large waves occur due to many reasons: storms, gravity, winds, sea surface tensions. A way the ocean systems form waves is from the fetch of a storm, which is the distance of the wind. Even places far from the reach of storm winds can be influenced from thousands of miles across the Pacific Ocean. The fetch of the storm can push a swell of wave trains far into the distance.

The distance from trough to crest on a wave is its height, H. The distance from crest to crest or trough to trough is the wavelength or lambda. The period of the wave is the time it takes for the wave to travel the wavelength. The velocity of the wave is the wavelength divided by the period.

$$\text{Power from waves}\left(\text{W}\right) = \sim 0.5 \times \text{Period}\left(\text{s}\right) \times \text{Height}\left(\text{m}\right)^2$$

The power capacity of the wave device is ½ times its height (meters) squared, multiplied by its period (seconds). This power, much like the wind power resource, is at the mid latitudes where the trade winds are reinforced with local winds. This explains why regions proposed for wave include the North Sea, Northern Pacific from California to Alaska, New Zealand, South Africa, and the Southern tip of South America. It is in these regions that average wave heights are highest and power potential is greatest.

5

Wave power devices can be placed in several different kinds of places. Some classifications of wave power devices are based on location: shoreline, nearshore, and off-shore devices. Shoreline wave devices are placed closer to land and have the advantage of being close to the utility network and are relatively easy to maintain. Nearshore wave devices can be placed in shallow water; a disadvantage is energy can be lost due to friction with the seabed. Offshore wave devices placed in deep water are tethered to the seabed and have much greater potential energy, but are difficult to construct, operate, and maintain.

WECs harness wave energy and with three different wave device types: point absorbers, attenuators, and terminators. In point absorbers, the WEC is similar to a buoy and moves vertically, up and down; it does not matter from where direct waves come from. Point absorbers are mostly used in nearshore wave locations and are non-directional devices (Rusu and Onea 2017). In attenuators, the WEC is oriented along the wave direction and generates energy as the wave passes through the system length; it appears to be on the water surface. The Salter duck, the Bristol cylinder, the Pelamis anaconda, and the clamshell are three specific types of attenuators. Terminator WECs are for onshore waves, also called as overtopping. The design acts as a breakwater—waves break on a ramp to be collected in a reservoir, driving a low-head turbine (Fig. 5.16).

The cost of wave technology can be quite expensive, since it is still in early development stages. The cost to build and operate wave devices can be very expensive. Bringing forward projects today will require patient partners, as well as subsidies, federal tax credits, grants, loans, support, and/or sponsorship from local government, alongside a lot more research. Costly underwater transmission cables will also be required to deliver power, so siting closer to shores lowers transmission line costs. One idea is to co-site wave power, wind turbines, or floato-voltaics to share infrastructure, minimize costs of underwater transmission lines, and reduce variability of output from changing wave and wind conditions (Rusu and Onea 2017).

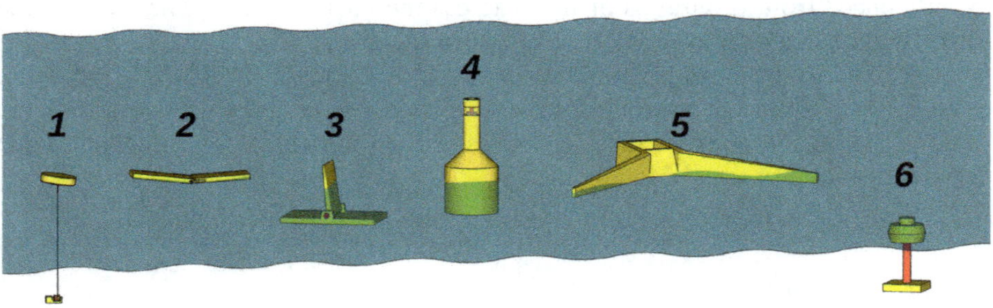

Fig. 5.16 Types and principles of wave power devices

Wave energy projects will have to account for disturbance or destruction to marine life, possible threat to navigation from collision, interference with commercial and sport-fishing, or degradation of ocean view from devices and transmission lines. Marine sanctuaries would be one of the locations to avoid attempting to build wave energy projects because they would face legal hurdles and regulatory constraints. Marine spatial planning with social scientists and biologists will be essential.

Improving wave condition forecasts will be critical to the success of wave power. Wave power can change depending on the season. In most places, the winter season generates more wave power than in summer. There are physics-based versus time series models that help predict waves. Physics models have been found to have good forecasting for longer horizons, and series model is better for short horizons (Reikard et al. 2015). Wind and wave weather conditions are closely related.

Even though wave power is a bit far from widespread deployment, it can provide a clean, reliable, and renewable source for some sites where the resources are good. Wave power can have minimal impacts to the environment, but there are still barriers with implementation such as cost and lack of infrastructure.

5.8 Tidal Power

Tidal power as the name implies harnesses the ebb and flow of the tides, as they move in and out. The main driver of this energy is the variations in gravitational pull mainly from the moon but also the sun, to make wake action occur in the world's oceans. Kinetic energy is harvested from the tides either from the flow through turbines in a specific direction or by impounding the tide behind a structure that is similar to a dam. Estimates for the resource potential vary, but it could be as high as 3 TW (compared to ~20 TW of average power demand).

Tidal stream generators are tidal devices that capture energy from a flow or current of water. As the tides pour into and out of estuaries and sloughs, there is usually a dominant direction that these devices are able to collect. A tidal barrage is more similar to hydroelectric facilities, in that the incoming tides are impounded behind the dam-like structure and flow through a penstock to turn a table. The Severn River is a famous site of a tidal barrage, proposed in 1921.

The power of a tidal basin and barrage can be estimated using the following equation, which represents the amount of power available from a specific-sized tidal basin with a given tidal range. Note that the tidal range is raised to the square, so the height of tidal range is a larger driver of power compared to the area of the tidal basin.

$$P_{\text{ave}} = \frac{\text{Density of fluid kg/m}^3 \times \text{Gravitational constant}\left(10\ \text{m/s}^2\right)\ \text{Area}\left(\text{m}^2\right) \times \text{Tidal range}\ (\text{m})^2}{2 \times \text{Time interval from high to low tide}\ (\text{s})}$$

Example 5.2 Estimate the Tidal Power Available in San Francisco Bay

$$P_{\text{ave}} = \frac{1{,}000\ \text{kg}/\text{m}^3 \times 10\ \text{m}/\text{s}^2 \times 1{,}000{,}000\ \text{m}^2 \times 1.5\ \text{m}^2}{2 \times 4 \times 10^4 \text{s}}$$
$$= 281\ \text{MW}$$

5

Tidal stream generators and tidal barrages offer some opportunities for renewable power in places where they may also help serve other kinds of infrastructure needs like flood control or to protect urban areas from storm surges. Overall, they offer limited opportunities just because the places where much of the power can be harnessed are both too few and often have conflicting uses. A tidal barrage on the Rance, a tidal estuary in Brittany, has operated for 50 years. The Severn Barrage proposed between Wales and England could provide 8 GW.

Early lessons can be gleaned from tidal generation developments on the Orkney Archipelago on the north coast of Scotland in an area off the North Sea. Key lessons show that places are undergoing development to harness these great seas and are also undergoing social changes. Understanding the social implications of marine energy more broadly includes these considerations related to cultural change and economic impacts (Conway et al. 2014).

References

Abbasi, S. A., & Abbasi, N. (2000). The likely adverse environmental impacts of renewable energy sources. *Applied Energy, 65*(1–4), 121–144.

Ballard, K. P., Everett, M. E. C., & Everett, W. C. (1991). Utilities and decommissioning costs: The meeting of technology and society. *The Energy Journal*, 29–41.

Behrsin, I. (2019). Green energy from garbage? A case study of municipal solid waste's contested inclusion in Maryland's renewable portfolio standard. *Case Studies in the Environment*. https://cse.ucpress.edu/content/early/2019/11/01/cse.2019.002048.

BNEF (Bloomberg New Energy Finance). (2018). World Reaches 1,000 GW of wind and solar, keeps going. https://about.bnef.com/blog/world-reaches-1000gw-wind-solar-keeps-going/.

Brumfiel, G., & Cyranoski, D. (2011). Quake sparks nuclear crisis. *Nature, 471*, 273–275.

California Energy Commission (2020a). Geothermal energy in California. California Energy Commission. www.energy.ca.gov/geothermal/background.html.

California Energy Commission (2020b). California geothermal energy statistics & data. California Energy Commission. www.energy.ca.gov/almanac/renewables_data/geothermal.

Conway, F., Watts, L., Colton, J., & Kerr, S. (2014). Establishing an agenda for social studies research in marine renewable energy. *Energy Policy, 67*, 694–702.

Crosby, A. W. (2006). *Children of the sun: A history of humanity's unappeasable appetite for energy*. New York: WW Norton.

Davis, L. W. (2012). Prospects for nuclear power. *Journal of Economic Perspectives, 26*(1), 49–66.

Fairhead, J., Leach, M., & Scoones, I. (2012). Green grabbing: A new appropriation of nature? *Journal of Peasant Studies, 39*(2), 237–261.

Firestone, J., Hirt, C., Bidwell, D., Gardner, M., & Dwyer, J. (2020). Offshore wind power siting: A mixed methods exploration of process fairness through the lens of trust, engagement, and individual involvement. *Energy Research and Social Sciences, 62*, 101393, accepted for publication.. https://doi.org/10.1016/j.erss.2019.101393.

Fraunhofer (2020) Photovoltaics Report. https://www.ise.fraunhofer.de/content/dam/ise/de/documents/publications/studies/Photovoltaics-Report.pdf.

Global Wind Energy Council. (2018). Global wind report 2018. https://gwec.net/global-wind-report-2018/

Hall, A., & Branford, S. (2012). Development, dams and Dilma: The saga of Belo Monte. *Critical Sociology, 38*(6), 851–862.

Hernandez, R. R., Hoffacker, M. K., Murphy-Mariscal, M. L., Wu, G. C., & Allen, M. F. (2015). Solar energy development impacts on land cover change and protected areas. *Proceedings of the National Academy of Sciences 112*(44), 13579–13584. https://doi.org/10.1073/pnas.1517656112.

IEA (2018). China. https://www.iea.org/countries/china.

Jager, H. I., Efroymson, R. A., Opperman, J. J., & Kelly, M. R. (2015). Spatial design principles for sustainable hydropower development in river basins. *Renewable and Sustainable Energy Reviews, 45*, 808–816.

Kaldellis, J. K., Apostolou, D., Kapsali, M., & Kondili, E. (2016). Environmental and social footprint of offshore wind energy. Comparison with onshore counterpart. *Renewable Energy, 92*, 543–556.

Kelly, S. (2019). Megawatts mask impacts: Small hydropower and knowledge politics in the Puelwillimapu, Southern Chile. *Energy Research & Social Science, 54*, 224–235.

Kessides, I. N. (2010). Nuclear power: Understanding the economic risks and uncertainties. *Energy Policy, 38*(8), 3849–3864.

Li, T. M. (2018). After the land grab: Infrastructural violence and the "Mafia System" in Indonesia's oil palm plantation zones. *Geoforum, 96*, 328–337.

Lovich, J. E., & Ennen, J. R. (2011). Wildlife conservation and solar energy development in the Desert Southwest, United States. *Bioscience, 61*(12), 982–992.

Murdoch, J. (1997). Towards a geography of heterogeneous associations. *Progress in Human Geography, 21*(3), 321–337.

Musial, W., Beiter, P., Schwabe, P., Tian, T., Stehly, T., Spitsen, P., et al. (2017). *2016 offshore wind technologies market report* No. NREL/TP-5000-68587; DOE/GO-102017-5031). NREL, Golden, CO, USA.

Pattey, E., Trzcinski, M. K., & Desjardins, R. L. (2005). Quantifying the reduction of greenhouse gas emissions as a result of composting dairy and beef cattle manure. *Nutrient Cycling in Agroecosystems, 72*(2), 173–187.

Perlin, J. (1999). From space to earth: the story of solar electricity. Earthscan, New York.

Rajaram, R., Siddiqui, F. Z., & Khan, M. E. (2011). *From landfill gas to energy: Technologies and challenges*. New York: CRC press.

Rampini, C. (2016). *Impacts of Hydropower Development along the Brahmaputra River in Northeast India on the Resilience of Downstream Communities to Climate Change Impacts* (Doctoral dissertation, UC Santa Cruz).

Randolph, J., & Masters, G. M. (2008). *Energy for sustainability: Technology, planning, policy*. Washington, D.C.: Island Press.

Reikard, G., Robertson, B., & Bidlot, J. R. (2015). Combining wave energy with wind and solar: Short-term forecasting. *Renewable Energy, 81*, 442–456.

Reisner, M. (1993). *Cadillac Desert: The American West and its disappearing water*. New York: Penguin.

Rusu, L., & Onea, F. (2017). The performance of some state-of-the-art wave energy converters in locations with the worldwide highest wave power. *Renewable and Sustainable Energy Reviews, 75*, 1348–1362.

Schramm, M. P., Bevelhimer, M. S., & DeRolph, C. R. (2016). A synthesis of environmental and recreational mitigation requirements at hydropower projects in the United States. *Environmental Science & Policy, 61*, 87–96.

Smil, V. (2004). World history and energy. *Encyclopedia of Energy, 6*, 549–561.

Srinivasan, T. N., & Rethinaraj, T. G. (2013). Fukushima and thereafter: Reassessment of risks of nuclear power. *Energy Policy, 52*, 726–736.

Trávníček, P., Kotek, L., Junga, P., Vítěz, T., Drápela, K., & Chovanec, J. (2018). Quantitative analyses of biogas plant accidents in Europe. *Renewable Energy, 122*, 89–97.

USDA (U.S. Department of Agriculture). (2015). Biogas opportunities roadmap progress report. DOE, USDA, EPA. https://www.rd.usda.gov/files/Biogas-Roadmap-Progress-Report-v12.pdf.

Visschers, V. H., & Siegrist, M. (2013). How a nuclear power plant accident influences acceptance of nuclear power: Results of a longitudinal study before and after the Fukushima disaster. *Risk Analysis: An International Journal, 33*(2), 333–347.

World Nuclear Association. (2019). Nuclear power in the world today. http://www.world-nuclear.org/information-library/current-and-future-generation/nuclear-power-in-the-world-today.aspx

Sustainable Energy Indicators

Contents

D. Mulvaney, *Sustainable Energy Transitions*,
https://doi.org/10.1007/978-3-030-48912-0_6

Learning Goals

Readers of this chapter will be presented with

- key concepts and problem sets related to sustainability;
- themes are concepts from industrial ecology;
- life-cycle assessment metrics and carbon and water footprint tools; and
- calculation of energy return on investment and energy payback time metrics.

Overview

This chapter introduces key concepts in sustainability science and engineering. The tools in this chapter include life-cycle assessment, industrial ecology/metabolism, risk and alternatives assessment, vulnerability analysis, and land-use science. Building on what readers have read in previous chapters about the fundamentals of energy science and their understanding of energy natural resources, this chapter offers opportunities to propose ways to evaluate impacts of energy technologies, infrastructure, and resources on Earth's socio-ecological systems.

An emerging field of study called sustainability studies can lend many useful tools to analyze energy systems (Clark and Dickson 2003). Different forms of environmental accounting have emerged with the proliferation of attempts to incorporate environmental externalities to activities in the productive economy. The following sections offer key concepts and analytical strategies to understand social and environmental dimensions of energy systems.

6.1 Industrial Ecology

Industrial ecology is an approach to rethinking production systems to minimize waste based on designs that mimic ecosystems, where wastes become feedstocks or inputs to other processes. Industrial ecology research is often exploratory and descriptive in nature, a metaphor for how production systems could be integrated (Ehrenfeld 2004). The underlying idea of industrial ecology is to move beyond end-of-pipe pollution solutions to radically rethinking production processes to prevent pollution and pursue environmentally benign production. The goal of this radical rethinking has led to the exploration of the concept of industrial symbiosis, where facilities relying on shared and interconnected resource would be co-located to maximize efficiencies and synergies.

Definition

Industrial ecology is a metaphor used to describe industrial systems that do not create waste because processes use waste as inputs or feedstocks to other industrial processes.

Industrial ecology research is often exploratory and descriptive in nature. The concept sets out a vision of symbiosis between production systems, with strategically co-located facilities sharing and interconnecting energy, resource, and waste flows to maximize efficiencies and minimize waste (Chertow 2008). Commoner's four laws of ecology make for a few rules of thumb to think about eco-industrial production: (1) everything must go somewhere, (2) everything is interconnected, (3) nature knows best, and (4) there are no free lunches in nature (Commoner 1971). These principles have been applied to several eco-industrial parks designed to synergistically interconnect industrial systems to take advantage of waste products and other shared resources (Erkman 1997). While the principles of industrial ecology sound like effective means to redesign industrial systems to be more sustainable, beyond the widely cited and unplanned example of Kalumbord, Denmark, and a handful of others, it remains more of an academic concept than a common practice (Gibbs and Deutz 2005). More recently, some of these concepts are linked to compatible ideas from biomimicry (Benyus 1997) and green chemistry (Iles 2008).

6.2 Sustainability Indicators

Global environmental change is pushing on planetary boundaries for some impacts, including climate, nitrogen loads, and biodiversity. The goal of developing sustainability indicators is to have quantitative measures that represent the impacts from human activities so that it is possible to understand the most holistic approach or to determine the least impactful process, or identify hot spots in a commodity chain for improvement.

Sustainability indicator research has taken off in recent years as corporate social responsibility (CSR) reporting has increased (a large portion of Fortune 500 companies report sustainability indicators to their shareholders and the public), and these kinds of metrics are integrated into public policy and public-facing disclosures about environmental performance. As sustainability indicators have grown in popularity, much of the conversation focuses now on how to harmonize and standardize approaches to environmental measurement (Hopwood 2009). Efforts are aimed at understanding how to standardize how sustainability is measured, enumerated, and codified into sets of performance metrics. Histories and institutional legacies can shape standardization and environmental accounting and reveal how they are socially constructed and embedded with assumptions about the world (Heiskanen 1999). Indicators can be the outcome of political negotiation as much as any reflection of materiality (Rametsteiner et al. 2011).

Nonetheless, they are what we have to measure, and that gives us some information. Leading organizations preparing sustainability information include the Global Reporting Initiative (GRI). Indicators include air quality emissions (criteria include air pollutants, heavy metals, volatile organic compounds), water use and disposal, greenhouse gases, nitrogen effluents, and literally hundreds of other indicators. The crucial consideration in the development of the indicator is getting the impact right in the numerator and the unit of activity or output right in the denominator (e.g., tons of sulfur dioxide per megajoule).

6.3 Carbon Footprints

Most human economic activities are associated with carbon emissions. The relative greenhouse gas (GHG) contribution of individuals from the industrialized nations (the OECD, or G20) are much larger than developing countries, although many of these countries are undergoing rapid emissions growth, and some countries need to still gain access to improved livelihoods, including modern energy services (i.e., solving the energy justice and energy poverty problem). There are large differences between different individuals—from the wealthiest to poorest in each country, or those with different cultural values or habits. While national-level GHG inventories may be important for developing national policies for carbon and renewable energy, GHG inventories of individuals give everyone an opportunity to know their relative contribution, and perhaps act on that information or understand where the structural changes are needed. A carbon footprint is a GHG inventory for an individual, product, or organization. It looks at the various individual consumption patterns and behaviors—commuting, heating and lighting a home, using appliances, and so on—and estimates the relative GHG contribution of each, and aggregates those into a total carbon footprint.

Example 6.1 Product Carbon Footprint

A major mobile phone manufacturer recently announced the amount of electricity used for manufacturing their latest smart phone. Their factories have reported a consumption of 37,000,000 watts with the operating capacity of 24 hours per day. (1) How much energy is used per year in kWh? (2) Assume the GHG intensity of electricity is 1.4 lbs. CO_2e/kWh. Calculate the annual GHG emissions related to manufacturing this phone in lbs. of CO_2. (3) If the manufacturer made 15 million phones in one year, what is the carbon footprint in lbs. CO_2 per smart phone?

(1) $$37{,}000{,}000\,\text{W} \times \frac{1\,\text{kW}}{1{,}000\,\text{W}} \times \frac{24\,\text{hours}}{\text{days}} \times \frac{365\,\text{days}}{\text{y}} = 324{,}000{,}000\frac{\text{kWh}}{\text{y}}$$

(2) $$359{,}000{,}000\frac{\text{kWf}}{\text{year}} \times 1.4\ \text{lbs.CO}_2\,/\,\text{kWh} = \sim 454{,}000{,}000\ \text{lbs.GHG}$$

(3) $$359{,}000{,}000\frac{\text{kWh}}{\text{year}} \times \frac{1\,\text{y}}{15{,}000{,}000\,\text{smart phones}} \times \frac{1.255\,\text{lbs CO}_2\text{e}}{\text{kWh}} = 30.05\frac{\text{lbs CO}_2\text{e}}{\text{smart phone}}$$

There are international standards for conducting carbon footprints such as the International Standards Organization (ISO 14046) and numerous reporting entities such as the Carbon Disclosure Project and Global Reporting Initiative. The standards usually set some of the key rules for making assumptions or making decisions about how to allocate impacts and emissions.

The global range for personal carbon footprints varies from 1 to 100 metric tons annually, and there are probably more severe high-use cases. The average American has a carbon footprint of about 24 metric tons of CO_2 annually. Climate scientists argue that for humans to keep emissions at a reasonably low amount of global warming, per capita annual global GHGs must average somewhere around two metric tons of annual CO_2.

Assignment 6.1 Personal Carbon Footprints

This exercise is intended to think about the way we measure and the assumptions we make when we do. A carbon footprint calculator is one way to measure the impacts of your consumption, often measuring GHG emissions, but also other impacts. In this exercise we will be comparing the questions and data collected from several carbon footprint calculators and unpacking how metrics like carbon footprints are comprised. The calculators are archived at www.dustinmulvaney.com/SET

Part 1: This assignment should take 2–3 hours, so leave plenty of time. In this assignment, you will explore several online carbon footprint calculators. Calculate your carbon footprint using each of the calculators. Convert all totals to metric tons of CO_2. Think about what they are asking, how your behavioral information is being compiled, the kinds of questions they ask, the assumptions that are made, and how accurately their methodology captures the GHG impact of your actions. Look around at the website. What does the organization do? What are the organization's motives?

Part 2: Enter your carbon footprint for each of the calculators mentioned later in the chapter in metric tons CO_2 per person. Note that some outputs will not be in metric tons per person, and other calculators may be based on households. For parameters considered, write down notes about the types of questions the calculators ask about (electricity, transportation, food, etc.). For observations, make notes on other aspects, user friendliness, goals of the organization, and so on. Is this carbon footprint calculator idea a useful tool? Why or why not? What are the trade-offs between keeping it simple and getting the numbers right?

Examples of LCA Objectives

1. To compare the energy used in industrial systems (called functional units);
2. To compare CO_2 emissions or global warming potential;
3. To compare water use, solid waste generation;
4. To compare energy paybacks among different energy sources and technologies;
5. To characterize trade-offs between scenarios;
6. To get absolute numbers, and relative percentages of energy/materials consumption
7. To produce a design tool that can roughly estimate some kinds of impacts; and
8. To regulate the carbon intensity of fuels.

There are many useful LCA resources on the Internet, including on the US EPA and NREL websites.

6.4 Life-Cycle Assessment

There is a growing trend toward measuring the environmental impacts of commodities and production systems, using life-cycle assessment (LCA) and using these metrics to inform the development of sustainability standards. This environmental accounting framework is transforming how some commodity systems are governed, as they become integral parts of policymaking and increasingly influence production processes. LCA is a tool for advancing key concepts in sustainability studies, green design, and industrial ecology. It is a quantitative approach used by practitioners to explore the material impacts of certain socio-ecological relations—the labor, the communities, and landscapes though are reduced to numbers, so it is important to keep data in context. LCA is a tool capable of measuring flow of materials and energy and is a key framework for a range of environmental accounting efforts from optimization questions in environmental studies to disclosure of product performance.

LCA appears to be an objective accounting framework on the surface and therefore apolitical, but it too requires acts of commensuration. LCA is a "cradle to grave" approach used to quantify and qualify carbon footprints, greenhouse gas inventories, toxics, and energy payback, among other impact categories. LCA emerged alongside the broadening of support for what is known as the "Limits to Growth" perspective from the early 1970s. The Coca-Cola Company set the standard for LCA in 1969 with a study to characterize the resource and energy dimensions of glass versus aluminum containers. Eventually, this led to European Commission issuing a Liquid Food Container Directive in 1985. The International Standards Origination (ISO) 14000 has developed a standardized LCA methodology.

The two types of key LCAs are attributional versus consequential. Attributional essentially inventories all materials and assigns them emissions, whereas consequential further asks what are the consequences of producing that product or passing a policy (versus some other product that no longer needed to be produced or that captures some indirect effect such as land-use change).

LCA distills impacts into a quantitative metric. For example, carbon footprint calculator reports its life-cycle carbon emissions in tons or pounds of carbon. Some transportation studies report costs on cost per distance as a proxy for energy use. More robust LCAs "normalize" the data based on several metrics so as to make sure that measurements computed using one set of units are not obscuring the results. Take the following example that looks at the relative carbon intensity of several kinds of foods and normalizes the data based on CO_2 per household, per dollar spent on food, per calories, and per kilogram (▫ Fig. 6.1).

6.4.1 Goal Definition and Scoping

When setting the goal and scope, it is important to think about motivations and data availability, and ensure precisely what you want to do—clearly defined at the onset. Here, it is critical to define terms, metrics, identify data sources, and sensitivity analysis, essentially planning out the full suite of tools that will be used in the

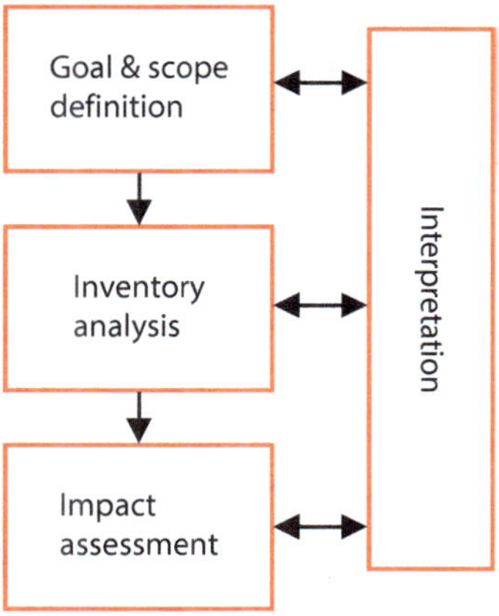

Fig. 6.1 The key steps in LCA briefly summarized in this chapter are iterative steps that always require thoughtful reflection and interpretation

analysis. Think about the outputs you want to see at the end: air pollutants, water effluents, hazardous emissions, and greenhouse gases. What are the chart and graphs, what are the categories or variables you would like to compare?

One important decision at this stage is setting the system boundaries. What are the end uses? Where does this product start and end? How far up the supply chain should we count? Which major operations should be included? How should co-products (when two or more products are produced from the same raw material or process) be assigned? What is the scope of geography? Where are the data from? National data? Regional data? Privately sourced data? One major challenge of LCA is the data vintage, which can be very old in some cases. Worse still is having no data at all and relying on some rules of thumb based on some of the basic ideas behind a process to use a proxy instead. For example, if the process is exothermic and requires heating, one could estimate this energy requirement. If the energy requirements to make a certain kind of plastic are unavailable, it may be suitable to use a similar product.

6.4.2 Life-Cycle Inventory

1. Life Cycle Inventory
   (a) The life-cycle inventory (LCI) is where the data are compiled about a particular project. Here, it is important to develop flow diagrams of the main system and subsystems, or the full industrial ecosystem, including some stages of the analysis if deemed important:
      (i) Raw material acquisition from Earth;
      (ii) Manufacture (material manufacture, fabrication, packaging);
      (iii) Use/reuse/maintenance (what it consumes/produces during use);
      (iv) Co-production (and justified means to allocate emissions, mass, or economic value are common); and
      (v) End-of-life (EOL) (will energy and emissions for managing EOL be added to the overall impacts, or will there be credit if there are materials recycled?).

   (b) Develop a data collection plan for an impact category:
      (i) Energy (Btu, MW), carbon? Heat (therms, ccf)? Global warming potential?
      (ii) Natural resources? raw materials? A toxic material? A rare or costly material?
      (iii) Green Job Creation? Economic Multipliers?
      (iv) Air pollution? Solid or Toxic Waste?
      (v) Prices/Costs/Feed-in-tariffs/Rebates?
      (vi) Broader concerns/Habitat/Safety/land use?
      (vii) Qualitative information, ethical concerns?
   (c) Collect data (what level of accuracy is needed?)
2. Life-Cycle Impact Assessment
   (a) Evaluate and characterize impact categories;
   (b) Weigh and emphasize most important aspects, identify values held by stakeholders; and
   (c) Identify most significant issues, evaluate, and document.
3. Life-cycle interpretation, conclusions, and recommendations for life-cycle management.

There is a growing trend toward measuring the environmental impacts of commodities and production systems using life-cycle assessment (LCA) and using these metrics to inform the development of sustainability standards. This environmental accounting framework is transforming how some commodity systems are governed, as they become integral parts of policymaking and increasingly influence production processes. Yet, while LCA is a critical tool for advancing key concepts in sustainability studies, green design, and industrial ecology, the quantitative approach used by its practitioners tends to erase the socio-ecological relations—the labor, the communities, and landscapes—that constitute these commodity chains. This means that frameworks like LCA have only limited purview into questions of the distribution of impacts—a question core to environmental justice. LCA-derived metrics can have an effective political agency, owing to their nature as numbers because numbers purport to be more "scientifically correct" (Porter 1995).

This means that frameworks like LCA have only limited purview into questions of the distribution of impacts—a question core to environmental justice. LCA-derived metrics can have an effective political agency, owing to their nature as numbers because numbers purport to be more "scientifically correct" (Porter 1995). This chapter draws on several life-cycle assessment and environmental justice case studies related to photovoltaic commodity chains, where LCA has become an industrial and policy norm and where a robust LCA practitioner community has emerged over 30 years, trying to shape the life-cycle sustainability of photovoltaics.

LCA tools and models in particular are said to have a central role, "not in solving problems, but constructing problems in a distinctive ways" (Heiskanen 2002, 434). As Stone has remarked, "counting requires judgment about inclusion and exclusion" (Stone 2002, 165). So the creation of any metric involves drawing boundaries around what should count, and what should not, or even how much they should count.

Impact categories used in LCA include global warming potential (CO_2e); stratospheric ozone depletion (CFCe); particulate matter ($PM_{2.5}$, PM_{10}); photochemical smog (non-methane volatile organic compounds—VOCs); eutrophication (NO_3e); terrestrial/aquatic ecotoxicity toxicity (fraction species affected); human toxicity (years life lost); acidification (SO_2e); abiotic resource depletion; and blue, green, and gray water footprints (Fthenakis et al. 2008).

Arguably, the weakest aspect of using LCA in policy is the subjective elements of the process throughout LCA—judgments about what to include, exclude, and how to measure. But it is the interpretation phase that is often most challenging because it relies on understanding the systems and how to put the LCA findings into context.

Assignment 6.2 Consequential LCA of a Photovoltaic Farm

A huge scale up of solar resources requires sustainability in production. In this assignment, a life-cycle assessment of a photovoltaic farm will be used to estimate the full greenhouse gas impacts of choosing between several different types of land choices.

Life-Cycle Inventory of Energy

	MJ/m² module
Silicon winning and purification	2,200
Wafer production	1,000
Cell/module processing	300
Module encapsulation materials	200
Overhead operations and equipment manufacture	500
Total w/o frame	*4,200*
Aluminum frame	400
Total module	*4,600*
Total balance of system	*750*
Carbon intensity of energy	50 grams CO_2e/MJ

Estimate the GHG emissions associated with making 10 km² of photovoltaics. Assume 80% of the area of photovoltaic module, the balance as roads and perimeter (in other words, the total area of the solar farm is 80% of each square kilometer.

The full square kilometer (km²) is still required for the overall project to work though (All are not covered with solar panels). Estimate next the land-use carbon intensity related to developing two different project sites. The first is a deciduous forest, where there are about 400 tons of CO_2e per km². The second is croplands that have 50 tons of CO_2e per km².

What are the total emissions associated with land-use change?

What are the total emissions of making a 10 km² photovoltaic farm, including the module and balance of systems?

How much energy does this photovoltaic farm generate over its lifetime in kWh?

If this solar farm is displacing natural gas of a carbon intensity of 0.5 kg of CO_2/kWh, how much carbon is displaced over its lifetime?

What is the GHG return on investment (ROI) for this project for each of the land use types?

6.5 Energy Return on Investment

A key metric used to understand the energy balance of different sources and technologies is energy return on investment (EROI). A useful analogy is a financial investment. If you invest a dollar and get US$5 in return, you have a return on investment of US$5. The EROI metric represents a multiplier that can be expected from an investment. Back to energy, EROI metrics are important because there can be hidden losses or inefficiencies not readily apparent. For example, 1 unit of energy yields 7 units of gasoline, whereas 1 unit of energy might only yield 3 units of ethanol. Or, worse, less than 1! Research on ethanol energy production finds that corn-to-ethanol EROI ranges between 0.75 and 1.7. This is largely due to the fertilizer used to grow the primary feedstock, corn, and the use of coal for heating in some facilities. Other crops could show higher EROIs on the order of ten. Beyond that, the limits of solar insolation make it challenging to get much more energy out. If land is a constraint, a photovoltaic system can generate more energy per acre.

The EROI of photovoltaic systems can range from 10 years to 22 years (Bhandari et al. 2015). A provocative research paper claims that the photovoltaic industry was a net electricity consumer until 2010 (Dale and Benson 2013). After that, the total energy return from photovoltaic systems is greater than was invested to make the devices. From here on, solar energy pays off GHG debts from manufacturing, supply chains, and other segments of the life cycle. There are some studies that claim the total EROI of the solar industry is now 2 or 3.

6.6 Energy Payback Time

The energy payback time (EPT) is meant to illustrate how long the energy investments in a renewable energy source take to pay off the initial or life-cycle energy investments. Solar PV modules typically have an energy payback, time to recover energy used in production and transport. The EPT for PV can range from six months as is claimed by some major photovoltaic manufacturers like First Solar to two and three years for crystalline silicon.

Definition

$$E = \left(E_{\text{mat}} + \right) / E_{\text{gen/yr}} - E_{\text{o\&m/yr}}$$

- E_{mat} = energy to extract, process, and transport raw materials to the manufacturing facility;
- E_{fab} = energy to fabricate a module from raw materials and transport it to the use site;
- E_{inst} = energy required for module installation;
- E_{elm} = energy required for end-of-life management;
- $E_{gen/yr}$ = energy generated by a module in one year; and
- $E_{o\&m/yr}$ = annual energy use for operation and maintenance

The above equation can help calculate the EPT for any particular source and technology. Analogously one could also use any other input or emission according to this idea and produce GHG payback time, and so on. In that concept, the accounting would be similar except that instead of energy investments along the life cycle, the analyst could measure GHG emissions input into the commodity system and time to breakeven with GHG savings.

GHG Emissions

GHG emissions from the life cycle of the PV system (g CO_2e),

- I = irradiation (kilowatt-hours per square meter per year [$kWh/m^2/yr$]),
- n = conversion efficiency, PR = performance ratio, LT = lifetime (years), and A = area of the module (m^2)

$$\text{GHGs} = (W)/(n \times \text{pr} \times I \times \text{LT} \times A)$$

6.7 Water Footprints

Data on water use and effluents will be used to develop blue, green, and water footprints and compare them to conventional production. Water footprints are indicators that represent impacts to water resources by estimating water quantities in addition to impacts on water quality. The approach divides water use into three components.

First, the blue water footprint estimates the volume of surface and groundwater withdrawn, consumed, or removed from the hydrologic cycle to produce a product. Since freshwater resources are scarce, this is often the most compelling and straightforward of the water footprinting techniques. It most accurately reflects the amount of water used.

Second, the gray water footprint is an indicator of water pollution, and it is defined as the amount of water required to assimilate water pollution to an acceptable threshold or standard (Mekonnen and Hoekstra 2011). The gray water estimation requires a two-step examination. The first step is to estimate a dilution factor, which requires to estimate some water quality threshold (effluent limitation or drinking water standard) that determines to what degree water can be contaminated. Second, it requires a measure or estimate to the degree of water contamination. With the knowledge of these two factors, we can estimate the dilution factor.

$$\text{Dilution factor} = \text{Contamination concentration/Water quality threshold}$$

With a dilution factor known, an estimate for the gray water footprint can be prepared by coupling information about the amount of water disposed of and normalized by the amount of natural gas production.

$$\text{Grey water footprint} = (\text{Water disposed} \times \text{dilution factor}) / \text{Natural gas production}$$

Finally, the green water footprint is the "use of the evaporative flow from the land surface" (Mekonnen and Hoekstra 2011, 1577). Green water footprints apply mainly to products produced in agriculture or forestry (Fig. 6.2).

Green water footprint is a measure of loss of water from transpiration during photosynthesis. It is primarily used for agricultural products and processes.

To get a full picture of water impacts water quality indicators are essential. Moreover, it makes a major difference whether the receiving water body is already impaired, protected, or requires some special consideration. Several key water indicators are (1) nitrate and phosphorous concentrate in the receiving river, stream, or lake, measured in milligrams per liter (mg/L). This is affected by crop choice, percent of residue and cover, the amount of nitrogen fertilizer applied; can cause eutrophication and hypoxia; and affects water drinkability, as nitrogen pollution is connected to issues affecting farmworker communities in the US, such as blue baby syndrome. Other key indicators are (2) suspended solids in streams (mg/L), which is also affected by crop choice, residue and coverage, or whether tillage is used. These effluents can cause benthic impacts and lead to siltation in water bodies. Finally, measuring for the presence of (3) herbicide/pesticide concentrations (mg/L) would help to also understand whether there is exposure to carcinogenic chemicals. Use of pesticides is also affected by crop choice, tillage, and the type and amount of herbicide/pesticide applied.

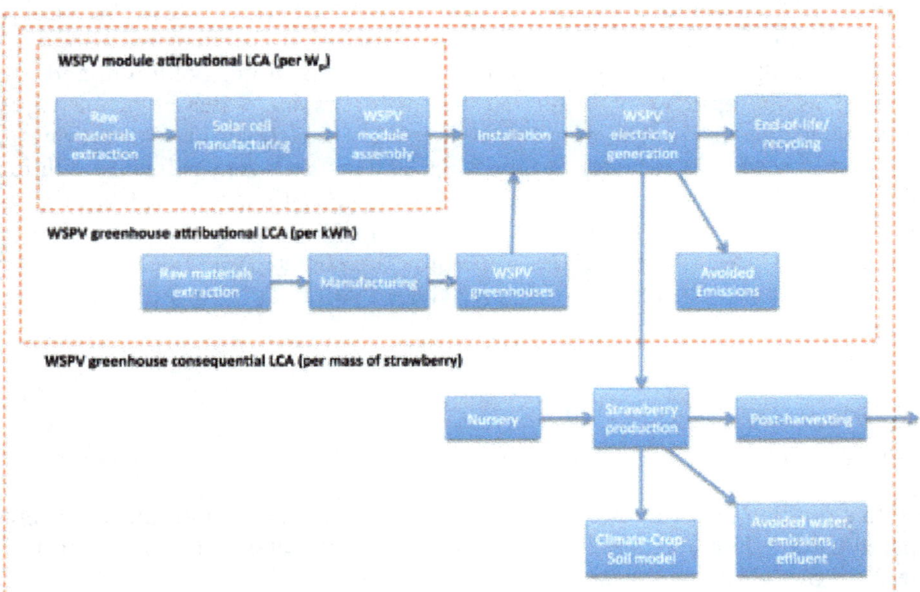

Fig. 6.2 LCA systems boundaries for an integrated solar greenhouse project

Assignment 6.3 Water Footprints for Natural Gas

In this assignment, we will estimate the amount of gray and blue water used to produce natural gas from shale of a specific well in Pennsylvania using the water footprint approach. Assume that the well requires 5 million gallons of freshwater and produces 15 million cubic feet of dry natural gas. The well also produces 3.5 million gallons of wastewater with barium contamination concentration of 25 grams per milliliter with a water quality threshold of 2.0 grams per milliliter. Estimate the blue and gray water footprints.

6.8 Cradle-to-Cradle Thinking

Understanding PV module supply chains is complicated by the fact that companies have different levels of vertical integration. Some PV manufacturers produce all stages from polysilicon to PV module, while for other manufacturers, polysilicon could be three tiers removed from the module's assembly. Non-silicon and non-semiconductor supply chains are similar across PV manufacturers. ◘ Figure 6.3 illustrates the crystalline silicon PV module supply chain.

Polysilicon is the feedstock that solar wafer manufacturers start with when melting to solar grade crystalline silicon. Globally, polysilicon manufacturing takes place in several locations across the US, China, Germany, South Korea, Japan, Malaysia, the Philippines, and India.

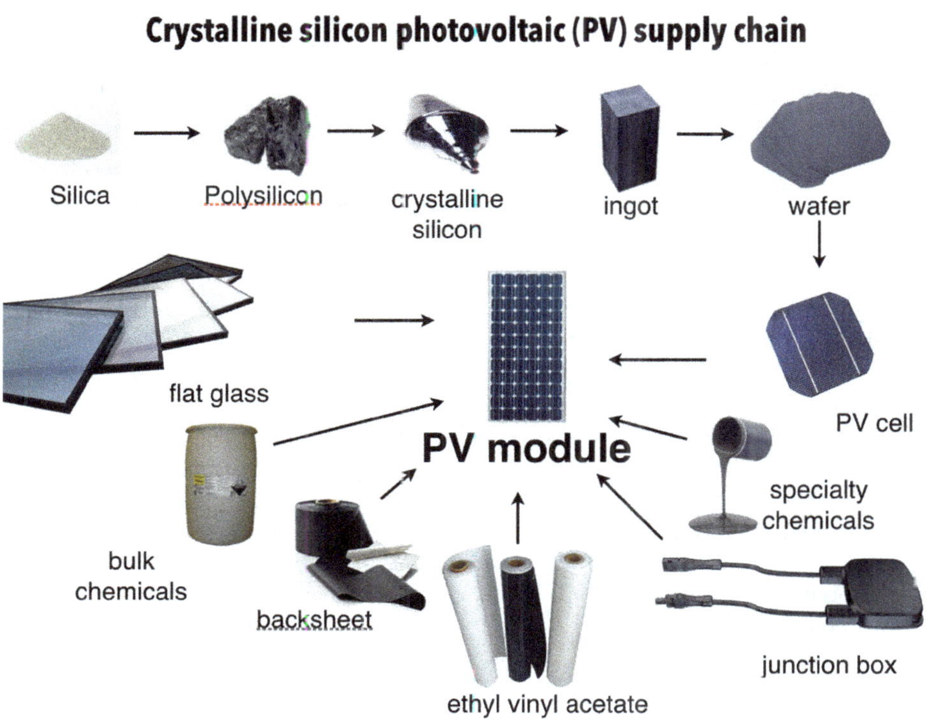

◘ **Fig. 6.3** Supply chain for photovoltaics

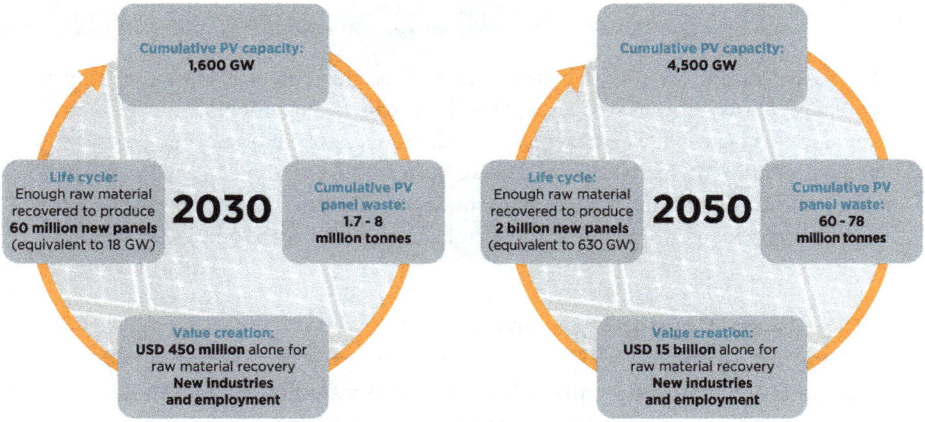

Fig. 6.4 Value potential from recycling end-of-life photovoltaics

In addition, several PV manufacturers make some of their own polysilicon. First-tier suppliers to polysilicon manufacturers are suppliers of metallurgical silicon, and second-tier suppliers are semiconductor-grade quartz mining operation (several wholesalers could be intermediaries between these purchases). Companies who buy polysilicon may sell wafers or carry out all of the subsequent steps of wafer, cell, and module manufacturing and assembly.

Thin-film PV modules are made from several building blocks. Thin-film PV manufacturers rely on specialty refineries that produce high-quality semiconductor powders of the metals they use in manufacturing. CdTe PV manufacturers require supplies of CdS and CdTe. Many thin-film PV technologies include indium (Fig. 6.4).

Assignment 6.4 Green Designs for Photovoltaics

Select a particular component in a photovoltaic module. Do some research on how that product is made, where it is made, and what kind of pollution hot spots or green design consideration is under investigation by scientists and researchers. The text below describes some of the other industries and companies in the supply chains for photovoltaics. Research what they do and what kinds of impacts are associated with these production facilities and sites of extraction.

Companies producing specialty chemicals are the suppliers of materials needed for cell preparation, manufacturing, and reactor cleaning among other plant operations. The materials used here include dopants, anti-reflective coatings, solvents, metallization pastes, and so on. Here, photovoltaic manufacturers could be buying from any number of Fortune 500 specialty chemical manufacturers.

Silver and aluminum metallization pastes are also key inputs provided by specialty chemical manufacturers. These pastes can contain lead and silver and are used in crystalline silicon PV modules.

Dopants are used to charge semiconductor materials and are supplied by other specialty companies.

Solder and ribbon are key pieces needed to interconnect a PV module.

Bulk chemicals are chemicals that are sold to multiple industries and are not specific to PV manufacturing. Hydrochloric acid is a major input to the manufacture of crystalline silicon PV modules.

The backsheet is a polymer material that protects PV cells. Backsheet suppliers include several large multinational chemical companies that specialize in advanced polymers such as Dow (US) and Dupont (US). Solar-grade flat glass providers to PV industry are flat glass manufacturers that are a subset of the global glass market.

Encapsulant is a polymer material affixed/cured atop the glass on a PV module to protect it from moisture.

Junction box The junction box is the interconnection interface between the electricity generated by the modules and the electricity system the power is being delivered to. It is usually on the back of the module. In most applications, the junction box is connected to an inverter, which is then routed to the electrical box in a home.

Frames Frames of aluminum are generically provided from aluminum suppliers. Many kinds of PV modules do not use frames.

What does a life-cycle perspective allow a designer to achieve? Think back the LCA assignment. Are there lessons learned from those production processes that might make good topics for designers to redesign.

Recycling, reuse, and dematerialization are design concepts that work against waste, resource depletion, and planned obsolescence.

Some design considerations include short-term usefulness, convenience, aesthetics, and whether it is made of reusable materials.

Moving past efficient design to being effective at achieving whatever the intended outcome is, whether it is GHGs or air pollution.

Assignment 6.5 Decision-Making Processes and Participatory Design of a PV Supply Chain

Today, people are asking questions about all sorts of products. People are shocked to learn, for example, about a critical mineral called coltan, which is used in every cell phone and is mined in artisanal systems in the People's Democratic Republic of the Congo. Similarly, cobalt is a critical metal in lithium-ion batteries. Think about an energy source or technology that will be important to a decarbonized future. Are there people affected by the production process? Find a material that will be key to that energy solution and trace its components and supply chains. Are there any pieces where things are currently in conflict?

- Cobalt in electric vehicle and stationary storage applications
- Manganese in lithium-ion batteries
- US midwestern corn for ethanol
- Brazilian soybeans for biodiesel

Think about these connections and tell a short story about that place from the Internet research you pursue. Are technologies neutral? Are they political? Are they just? Think about why should we care about the how a technology impacts us and others?

What if these communities that supply key inputs were asked about benefits from energy infrastructure and devices directly? What future do they see, and would they like to see the benefits their families and communities would reap? What would these communities invest in if they directed those investment decisions?

6

6.9 The Transition to Sustainable Industry

At the advent of the twenty-first century, production is increasingly prominent as a site of would-be policy governance and industry activity, compared to plant regulation. Proponents of this approach cite two reasons. First, production is associated with "unsustainable" environmental impacts. Many products are recognized as containing toxic, persistent, or bioaccumulative chemicals. Their use, storage, transport, and deterioration over time, not just manufacturing or disposal, can lead to dissipation into the environment. By targeting how products are chosen and made, a troublesome source of contamination can be addressed at the outset rather than at the tailpipe. This view seems to prevail in the US to date. Second, some think that focusing on production would promise some fundamental "solutions" to environmental problems. If industry is to attain greater sustainability, it needs to reconsider its production choices. By targeting the materials, water, energy, ecological, and social impacts of production at the stage of making design choices, it may be feasible to redesign.

Despite the advances in developing principals of green/sustainable chemistry during the 1990s, among other developments, chemical production continues to be a confounding area for industry, citizens, and governments in Europe and the US (Iles 2008). The EU has embraced "integrated product policy" and screening of chemicals, and the phase-out of some hazardous chemicals. US state and federal governments do not have a production policy, though they have encouraged what they call "product stewardship," regulatory exemptions aimed at product design, and voluntary tools. Generally, governments are reluctant to oversee production because of the diverse products and processes, and because this would affect the power of industry.

Most analysts of production policies in the US and the EU have believed that engagement by citizens, consumers, and environmental groups in product and process design are largely not feasible, so that it is for chemists and other technically trained actors to serve as designers and builders. Therefore, industry may face less scrutiny of its production practices, including adoption of practices that seek inherently safer process design (Mulholland et al. 2000). Research from science and technology studies (STS), anthropology, and sociology reveals that designers of products, processes, and buildings do increasingly make choices with environmental impacts, based on their beliefs about what users want, what regulatory systems oblige them to do, and the needs of engineering reliability (Guy and Shove 2000). There are many signs of lay actors in the US and Europe starting to exercise power in using scientific and technical knowledge to critique products, suggesting that citizens may help shape production choices.

Most important, policymaking and company practices appear not to have probed what "sustainability" might mean and how to make production choices transparent, traceable, and accountable. Frequently, the chemical industry claims that production is "sustainable" when they may not be under some interpretations. Companies may make technology and material choices because of their differing views of what counts as "green" or "sustainable." The chemical industry has already introduced a number of reaction pathways that reduce the use of toxic substances, but that introduce other toxic substances, generate vast amounts of wastes, and consume significant energy. Policies, industry practices, and incentives often do not take account of these features.

National Standards Foundation International Sustainability Leadership Standard for Photovoltaics is a case study of the process of stakeholder participation to develop sustainability standards that will allow students to understand the process of holding collaborative meetings and discussions about what the content of sustainability standards should be. Students will role-play different stakeholders, understand key issues, and collaboratively design a mock standard, before mastering the content of the actual standard.

6.10 Green Chemistry

Green chemistry is an area of research in sustainability studies that is being pursued by chemical industries and product designers. The chemical industry manufactures over 80,000 chemicals worldwide. Some of these chemicals are products in their own right, being sold on the market in the form of cleaning chemicals, pesticides, or dyes. Most chemicals, though, end up being incorporated into products, such as toys or cosmetics, that "downstream" companies in other industrial sectors make for consumption. In energy systems, there are chemicals used to extract, clean, and process fossil fuels. There are also chemicals used to manufacture photovoltaics and wind farm devices. In the US and in some parts of Europe, green chemistry is being touted as an effort to ameliorate the hazards of chemical manufacturing. Traditionally, production yield, profitability, product specifications desired by end users, and investment costs have driven product designers. But Anastas and Williamson, EPA scientists, argue: "Through careful planning, chemists are able to manipulate molecules and develop products and production that minimize harmful effects to human health and the environment" (Anastas et al. 2000). In Europe, scientists and industry also contend that sustainable chemistry should be the goal, instead of simply reducing the hazardous effects of chemicals. Since the early 1990s, a growing number of books on green chemistry have been published, not just covering the technical aspects, but looking to societal and policy implications.

Green/sustainable chemistry includes numerous, often uncommercialized, techniques and areas of applications. These range from designer catalysts that allow more efficient yields; changing the reaction conditions to use less heat and pressure; altering the composition of the feedstocks or reactants; finding solvents that avoid use of volatile organic compounds; and the use of renewable energy. This project focuses on several relatively well-established areas of the burgeoning field that are already being used by industry (with ample documentation), that have attracted political and regulatory interest in either or both of the US and Europe, and that have important human health implications.

6.11 Political-Industrial Ecology

The idea of a political-industrial ecology emerged out of geographic scholarship (Cousins and Newell 2015) (Fig. 6.5).

LCA as a practice has become deeply instilled in major institutions of environmental governance. Practitioners have developed guidance through the International Standards Organization on how LCAs should be conducted and what kinds of claims manufacturers and experts can make based on the data quality and other interpretive criteria. The International Energy Agency through working groups with the most prominent LCA practitioners has developed more detailed guidance on how to conduct this research on photovoltaics. In the peer-

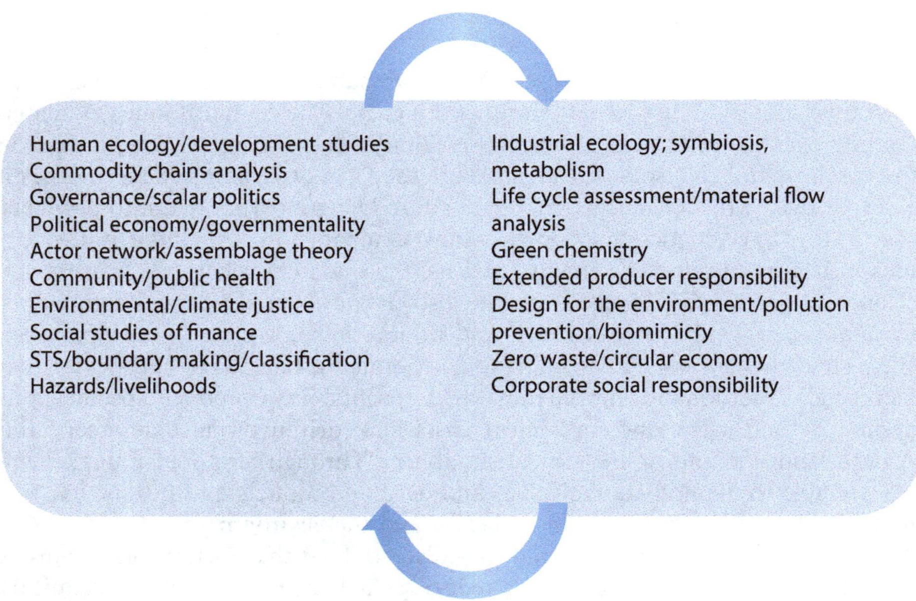

Fig. 6.5 Industrial-Political Ecology integrates social science research with the study of policy frameworks, normative goals, and tools used in sustainability science and ecological modernization discourse

reviewed literature, LCA is extensively used to compare energy systems. The systematic and quantitative approach of LCA readily lends itself to the accounting for environmental impacts of energy because energy resources can be easily made commensurable (Espeland and Stevens 1998). Energy is a derived demand—meaning the demand for energy services mediates demand for energy resources. The electricity grid requires managing a complex system of power plants and power demands drawing from various resources, making such comparisons about electricity resources not only possible but also meaningful for environmental and climate policy. People want electricity outlets to function and automobiles to move along; they do not, otherwise, care what resources make this possible. This means that energy lends itself to commensurable functional units; all energy can be converted to other units used to represent energy (kilowatt-hours, megajoules, barrels of oil equivalent, etc.). There are more LCAs written in English about energy systems than any other industry or sector. Database searches in April 2016 revealed 104,000 results for "life-cycle assessment" plus "energy" in Google Scholar and the same search in Science Direct yielded 115,710; similar searches for "life-cycle assessment" and "food" (89,160 Science Direct) and "agriculture" (38,900 Google Scholar; 31,185 Science Direct) yielded similar results. There is arguably no comparable set of industry-specific LCAs undergoing extensive harmonization, and LCA is also becoming embedded in many regulatory instruments, which in the US includes LCA-based performance standards set by the Environmental Protection Agency and the Department of Energy.

Conducting an LCA requires proceeding along several major stages, though, in practice, these are iterative processes, with constant tunings, adjustments, and quests for more representative data and model treatments. The first stage is to establish the goal and scope of the LCA. This is where the practitioner or client determines the major parameters of the life cycle, which ensures the approach meet the project objectives and expectations by confirming the functional unit, setting system boundaries, and identifying environmental impact categories. These impact categories include global warming potential, eutrophication, acidification, toxicity, ground-level ozone/smog formation, stratospheric ozone depletion, natural resource depletion, water pollution, water use, eco-toxicity, and human toxicity. There are a few organizations that standardize impact category data, including TRACI. A review of 150 published LCAs found over 90 different metrics reported in publications from 1976 to 2011. The proliferation of metrics has led to calls for greater standardization and harmonization. The National Renewable Energy Laboratory's (NREL) harmonization project produced a meta-analysis of energy-related LCAs (Heath and Mann 2012).

Environmental metrics, because they carry with them the power of numbers, have discursive power because they appear as impartial representations, distancing the subjective from the political. Relying primarily on quantitative data, LCA appears objective, overcoming problems of distance and distrust (Porter 1995). LCA is conducted in four phases: (1) goal and scope, (2) inventory, (3) assessment, and (4) interpretation. Note that despite LCA being described as a quantitative tool, two of the major phases are deeply qualitative and, therefore, mediated by human subjectivity. Even where objective accounting of inventories is done, the processes of translation require commensuration—reducing complex socio-ecological interactions to metrics. As Espeland and Stevens (1998) note, commensuration has a long history as "an instrument of social thought, and as a mode of power." Commensuration occurs when the complexities of social life and exchange are reduced to metrics.

Metrics can instill objects with political agency and semiotic power because numbers are mobilized for claims-making. As materialities become represented through LCA metrics, production systems can be detached from place and context, comparable with other systems in other parts of the world. Numbers appear to be immutable, undistorted, and easily comparable representations of facts. Numbers displace human agency in decision-making as they operationalize certain outcomes. LCA tools have a central role, "not in solving problems, but constructing problems in a distinctive way and making people aware of them" (Heiskanen 2002, 434). LCA is also limited to issues described and compared by the numbers, when not all key socio-ecological phenomena are easily reduced.

LCA began as a practice used to study material and energy flows since the 1960s. Initially used for internal environmental performance accounting in technology development and materials selection, LCA evolved into academic research clusters in private practitioners producing product LCAs for various reasons, including advertising, product design, and regulatory compliance. Heiskanen (2002) characterizes LCA as either representational or instrumental, which would have to do with the projects' goal versus scope. A motivation to conduct an LCA may be to capture existing performance or to study the consequences of changes to the product or process design. LCA is also used in several different energy regulatory instruments, including the US EPA's renewable fuel standard and California's low-carbon fuel standard. The Department of Energy's loan guarantee program that awarded US$60 billion in loans and grants through the American Recovery and Reinvestment Act required in LCA for any of its pre-commercial clean technology projects. LCA is also increasingly used in other environmental standards such as the Leadership in Energy and Environmental Design (LEED) certification for green buildings, where LCA concepts were introduced to the standard in 2010.

The calculative practices of LCA can be described in similar ways to the processes of commensuration and valuation/marketization. Calculating a value for nature and ecosystem services (Robertson 2012) and making markets (MacKenzie 2009) might work for economists looking to put a value on things like ecosystems services, but in reality, some things are more and less quantifiable.

Metrics delineate lines between expert and lay, creating artificial divisions between questions for technocrats and those for the rest of society. "The apparent objectivity of numbers, and of those who fabricate and manipulate them, helps configure the respective boundaries of the political and the technical" (Rose 1999, 199). Stone has remarked on how counting requires the creation of categories and how the creation of categories means leaving some phenomenon inside the category, and moving some outside. In Stone's own words, "counting requires judgment about inclusion and exclusion" (2002, p. 165). This is an important consideration in LCA, where setting the system boundary is an early stage of the analysis.

After the quantitative inventorying and impact analysis in LCA is completed, the analyst is required to interpret the results. How can LCA be used to help understand just and sustainable low-carbon energy transition without reducing materiality and lived experiences to numbers? Social life-cycle assessment has emerged as a subtheme in LCA and could be a core research topic for an industrial-political ecology.

The mixed methods used in commodity chain analysis and LCA can be integrated to better understand the justice and sustainability dimensions of solar energy transitions. In 2015, global use of solar energy was nearly 2%, but to make meaningful contributions to future energy demands, the solar industry will need to scale production to fill in 50% or greater of global electricity supplies. This means rapid scaling of natural resource production, land change, and employment in these sectors.

This research attempted to sketch these frameworks side by side to look for contradictions related to impacts or areas of emphasis, and quality of data. In both the LCA and commodity chain frameworks, there were gaps in data quality, owing to proprietary and privacy considerations. Nonetheless, several key conclusions can be reached based on information these approaches offer. The first point is similar to a conclusion that LCAs illustrate only marginal differences among photovoltaic technologies, but they articulate significant differences when comparing these impacts to other energy sources. Studies consistently find considerably more GHGs and other pollutants from fossil fuels compared to solar energy.

The harmonization research conducted by the team at NREL has been helpful at shedding light on solar energy production. One challenge is that LCAs are rarely representative of the entire fleet of production nor are they based on all manufacturing processes. For example, most photovoltaic manufacturing today is in Asia, yet fewer than ten LCAs (< 5%) are based on manufacturing there. Other issues relate to the vintage of the data. LCA can take several years to produce, and by the time it is published, the data can be 5–10 years old already, or even older.

Argonne National Laboratories conducted an LCA to attempt to compare the carbon footprint of PV modules made in Europe versus those made in China and found that the high-carbon intensity of energy in China led to double the carbon footprint. Yet, this does not account for PV manufacturers that may utilize onsite-distributed energy such as has been reported by SunPower, First Solar, and SolarWorld. Sustainability, in practice, is changing organizational cultures and practices, and this is true with solar energy companies.

LCA alone is unable to reconcile concepts of resilience and vulnerability insofar as it is using these to define goals and scopes and interpret results. Our tendency to value numeracy and more scientifically valid (even if it is not) plays a role in the effectiveness of sustainability discourse around sustainability metrics. Analysts need to keep a critical eye toward sustainability metrics, while also utilizing and learning from them to improve production processes that can drive production practices that do embody to core elements of sustainability. LCA is helpful in industrial ecology efforts as it can map out different ways that production systems can be put together.

Political-industrial ecology is a concept meant to convey the network of concepts, ideas, substances, and places that constitute production-consumption linkages

(Cousins and Newell 2015). This broadly can include everything from the subjectivities of consumers to the landscapes, communities, and sites where raw materials are extracted.

A previous era of semiconductor industries making similar clean claims—no smokestacks, clean rooms, and green manicured campuses—left legacies of environmental injustices from Silicon Valley California to factories in East Asia. The similarities between electronics and photovoltaics manufacturing is a common point of reference threaded throughout the peer-reviewed research on environmental health and safety of photovoltaics. While LCA does account for this chemical use, it offers little insight into systems of production where exposures occur. Qualitative tools and LCA can be brought together to understand the implications of scaling photovoltaic production to aid socio-technical transitions to solar power. Investments in research and development and policy incentives can better emphasize innovations that not only reduce the cost of photovoltaics but harness opportunities to produce photovoltaics in socially beneficial and environmentally benign ways. Yet, today, LCA remains mainly a tool to validate existing socio-technical performance, not to identify opportunities to improve environmental performance, although some manufacturers report that LCA information does feed into decisions to change chemical composition. CSR efforts alone should not be confused with advancing a just and sustainable world. There is much that can be learned by bringing the tools of critical social science research with the material understandings of production systems political-industrial ecology can learn from the outputs of life-cycle assessment, and vice versa.

References

Anastas, P. T., Bickart, P. H., & Kirchhoff, M. M. (2000). *Designing safer polymers*. New York: Wiley.

Benyus, J. M. (1997). *Biomimicry: Innovation inspired by nature*. New York: Harper Perennial.

Bhandari, K. P., Collier, J. M., & Ellingson, R. J. (2015). Energy payback time (EPBT) and energy return on energy invested (EROI) of solar photovoltaic systems: A systematic review and meta-analysis. *Renewable and Sustainable Energy Reviews, 47*, 133–141. https://doi.org/10.1016/j.rser.2015.02.057.

Chertow, M. R. (2008). Industrial ecology in a developing context. In *Sustainable development and environmental management* (pp. 335–349). Dordrecht: Springer.

Clark, W. C., & Dickson, N. M. (2003). Sustainability science: The emerging research program. *Proceedings of the National Academy of Sciences, 100*, 8059–8061. https://doi.org/10.1073/pnas.1231333100.

Commoner, B. (1971). *The closing circle* (Vol. 8, pp. 341–351). New York: Alfred A. Knopf. Cons, JD, & Hayes.

Cousins, J. J., & Newell, J. P. (2015). A political-industrial ecology of water supply infrastructure for Los Angeles. *Geoforum, 58*, 38–50. https://doi.org/10.1016/j.geoforum.2014.10.011.

Dale, M., & Benson, S. M. (2013). Energy balance of the global photovoltaic (PV) industry-is the PV industry a net electricity producer? *Environmental Science & Technology, 47*(7), 3482–3489.

Ehrenfeld, J. (2004). Industrial ecology: A new field or only a metaphor? *Journal of Cleaner Production, 12*(8–10), 825–831.

Erkman, S. (1997). Industrial ecology: An historical view. *Journal of Cleaner Production, 5*, 1–10. https://doi.org/10.1016/s0959-6526(97)00003-6.

Espeland, W. N., & Stevens, M. L. (1998). Commensuration as a social process. *Annual Review of Sociology, 24*(1), 313–343.
Frischknecht, R., Heath, G., Raugei, M., Sinha, P., & de Wild-Scholten, M. (2016). *Methodology guidelines on life cycle assessment of photovoltaic electricity* (No. NREL/TP-6A20-65291). Golden: National Renewable Energy Lab.
Fthenakis, V. M., Kim, H. C., & Alsema, E. (2008). Emissions from photovoltaic life cycles. *Environmental Science & Technology, 42*(5), 2168–2174.
Gibbs, D., & Deutz, P. (2005). Implementing industrial ecology? Planning for eco-industrial parks in the U.S. *Geoforum, 36*, 452–464. https://doi.org/10.1016/j.geoforum.2004.07.009.
Guy, S., & Shove, E. (2000). *A sociology of energy. Buildings and the environment: Constructing knowledge, designing practice*. London: Routledge.
Heath, G. A., & Mann, M. K. (2012). Background and reflections on the life cycle assessment harmonization project. *Journal of Industrial Ecology, 16*, S8–S11.
Heidegger, M. (1954). The question concerning technology. *Technology and Values: Essential Readings, 99*, 99–113.
Heiskanen, E. (1999). Every product casts a shadow: But can we see it, and can we act on it? *Environmental Science & Policy, 2*(1), 61–74.
Heiskanen, E. (2002). The institutional logic of life cycle thinking. *Journal of Cleaner Production, 10*(5), 427–437.
Hopwood, A. G. (2009). Accounting and the environment. *Accounting, Organizations and Society, 34*(3–4), 433–439.
Iles, A. (2008). Shifting to green chemistry: The need for innovations in sustainability marketing. *Business Strategy and the Environment, 17*(8), 524–535.
Leigh, M., & Li, X. (2015). Industrial ecology, industrial symbiosis and supply chain environmental sustainability: A case study of a large UK distributor. *Journal of Cleaner Production, 106*, 632–643. https://doi.org/10.1016/j.jclepro.2014.09.022.
MacKenzie, D. (2009). Making things the same: Gases, emission rights and the politics of carbon markets. *Accounting, Organizations and Society, 34*(3–4), 440–455.
Mekonnen, M. M., & Hoekstra, A. Y. (2011). The green, blue and grey water footprint of crops and derived crop products. *Hydrology and Earth System Sciences, 15*(5), 1577.
Mulholland, K. L., Sylvester, R. W., & Dyer, J. A. (2000). Sustainability: Waste minimization, green chemistry and inherently safer processing. *Environmental Progress, 19*(4), 260–268.
Poliakoff, M., Fitzpatrick, J. M., Farren, T. R., & Anastas, P. T. (2002). Green chemistry: Science and politics of change. *Science, 297*(5582), 807–810.
Porter, T. M. (1995). *Trust in numbers*. Princeton: Princeton University Press.
Rametsteiner, E., Pülzl, H., Alkan-Olsson, J., & Frederiksen, P. (2011). Sustainability indicator development—Science or political negotiation? *Ecological Indicators, 11*(1), 61–70.
Robertson, M. (2012). Measurement and alienation: Making a world of ecosystem services. *Transactions of the Institute of British Geographers, 37*(3), 386–401.
Rose, N. (1999). *Powers of freedom. Reframing political thought*. Cambridge: Cambridge University Press.
Stone, D. (2002). *Policy paradox*. New York: WW Norton & Company.

Low-Carbon Electricity Systems

Contents

D. Mulvaney, *Sustainable Energy Transitions*,
https://doi.org/10.1007/978-3-030-48912-0_7

Learning Goals

At the end of this chapter, readers should be able to:

- understand the connection between natural resources and final fuels;
- estimate the economic costs and value of different energy sources; and
- assess the relative impacts of different kinds of electricity system design.

Overview

Shifting from stocks of finite resources to flows of renewables can minimize many of the major impacts from energy systems. This chapter reviews the key concepts to evaluate electricity systems on energy production, economics, and minimize their impacts. Students will learn the geographies of renewable energy resources, supply chains, and generation capacity. A section on harvesting waste energy will include case studies of challenges where contaminants are present, environmental justice issues, and debates about whether waste incineration should be considered a renewable energy under portfolio standards in the US or feed-in-tariffs in Europe (or sui generis regulatory scheme). Concepts and techniques to evaluate renewables, including energy pay-back time, energy return on investment, levelized cost of electricity, are introduced, and the chapter has problem sets and exercises based on publicly available computer simulation tools developed by US federal research laboratories such as GREET (Wang 2005), the PV Watts calculator, and the System Advisor Model developed by the National Renewable Energy Lab and Argonne National Labs (NREL 2018).

Who gets to define, develop, and benefit from green electricity or transportation or industry? Answers to these questions are not entirely straightforward as technologies are entangled in politics, culture, and the economy, in addition to the biophysical systems that support human civilization and planetary ecosystems. Embracing sustainable energy transitions means evaluating the technologies, potential configurations, and understanding how they may shape the society.

7.1 The Electricity Grid System

Electricity grids are the largest machines in the world. This section briefly describes their organization and several key features. The first aspect to consider is that electricity systems are a natural monopoly. To avoid multiple transmission and distribution lines running in parallel to households that are being serviced which might be on the same block, jurisdictions grant utilities the right to entire markets. While, in the past, many utilities owned both the power plants and the systems of wires that delivered power to homes, systems evolved and developed, and utility deregulation opened new competitive markets for electricity generation. In some places, electric utilities have since shed their power plants and now contract with merchant power plants to acquire most of their electric power.

Definition

A grid is a collection of components and devices that transmit, distribute, and deliver electricity, including the transmission wires and towers, distribution wires and towers, substations, transformers, meters, and other essential parts, sometimes including the power generators, companies, and workers.

Physically speaking, an electricity grid is constituted by power plants, transmission and distribution infrastructures, and substations. Power plants generate alternating current (AC) electricity, which is stepped up at a nearby substation to higher voltages using equipment called transformers. This high voltage (100s of kilovolts or a few common ones are 230,000 volts; 345,000 volts; 440,000 volts; 500,000 volts; 750,000 volts) makes it easier to push significant amounts of power in a relatively thin wire. Moving significant quantities of electricity at lower voltages would require very thick wires otherwise. Remember voltage is the "push" or "pressure" of electrify energy (Fig. 7.1).

Most high-voltage transmission lines deliver AC current, although there are some high-voltage direct current (DC) transmission lines. The advantage of AC over DC is that AC power can have many taps to deliver locally into distribution circuits. DC lines may become more popular to move renewables to load centers. Once power reaches a distribution substation, it is stepped down to several hundred volts and carried on power poles. These are the familiar wooden or cement poles that distribute electricity in a neighborhood. As it approaches homes and businesses, a line-pole transformer steps down the voltage one last time, ready to be

Fig. 7.1 An electricity substation near Amboy Crater, California, in the Mojave Desert

delivered to the circuit breaker box in a home at 220 volts, where most outlets are 110–220 volts. This varies by region. In the US, most appliances and devices are designed to handle 110 volts, whereas European appliances and devices can withstand 220 volts. This is why travelers bring adapters for their equipment. It is not just a different plug socket, it is a different electric juice.

> **Social Discount Rates**
> The basic idea behind a discount rate is that there is a time preference on money. Most investors see the same fixed amount of money as less valuable in the future than it is today. The social discount rate is a measure of how much less valuable money is to an investor over time. The higher the discount rate, the faster that money declines in value in time. Environmental ethicists and practitioners have noted that the time preference on money can vary between private that highly discount the future and public investors that may not discount the future value of money as much and prefer a low discount rate (Howarth 2005). Most health and climate change–related investments have low discount rates of 3% (Smith and Haigler 2008).

In the US, the electrical grid—the largest in the world—can be divided into three sections. The Western US is the Western Interconnection (WECC), the *Electric Reliability Council of Texas* (ERCOT) covers Texas, and the East, Midwest, and South comprise the Eastern Interconnection. Some of these are integrated at transmission with Mexico and Canada. China has two major electric grids, and several smaller ones, but plans to interconnect them in the future. India's electric grid does not reach the entire people there and is not interconnected. In Australia, the transmission is split into two grids. Europe is mostly powered by the synchronous grid of Continental Europe. There are many energy analysts arguing for more grid integration to wheel renewable electricity across wider regions. In poorer parts of the world, there is limited access to the grid. National grids are just starting to roll out across much of Africa, but there is still a lack of access for the majority of people there.

The variability of demand and supply across a 24-hour cycle has historically followed a pattern that allowed power supply to be characterized as baseload power, peak power, and shoulder power. Baseload to some is already an anachronistic term, but it, generally, describes a power plant that runs most, if not all, hours per day. This means it has a high capacity factor. Peak power as it sounds is the power required to "top off the grid" at peak. This typically occurs late in the day when everyone is using power, but peak grid power shifts further into the evening as more and more solar and batteries are installed in the distribution network. This solar or storage displaces the peak from the midday (when it is sunniest) until late in the day when the sun is low in the sky (◘ Fig. 7.2).

7.2 Levelized Cost of Electricity

What is the levelized costs of electricity (LCOE) of an electricity generator energy compared to electricity from the grid? LCOE is a standard economic metric, representing the cost of electricity and allows for comparing energy sources with varying capacity factors or those with incommensurate operating and long-term

Fig. 7.2 Siting transmission infrastructure can pose its own challenges especially in high fire-risk or wilderness areas

fuel versus upfront capital costs such as with fossil fuel versus renewable generation sources (Branker et al. 2011). The main drivers of the LCOE calculation are capital costs, fuel costs, annualized operating and maintenance costs, and financing, as well as a capital recovery factor based on specified discount rate and power plant life (Timilsina et al. 2012) (Fig. 7.3).

Definition

Levelized cost of electricity is the cost of delivering electricity over some time period. It is commonly measured as cost per unit energy (US$ per kWh).

LCOE is commonly used to compare distributed to utility-scale solar. A sensitivity analysis can explore different discount rates and interest rates, and under different solar net metering tariff scenarios, including retail-rate net metering, net metering with fixed charges, avoided cost net metering, and other rapidly evolving policies (Deshmukh et al. 2012). LCOE sensitivities to federal, state, and utility incentives are also evaluated to understand the influence of these public policies on overall costs. Public policies considered in the scenarios include the US Investment Tax Credit and the US Department of Agriculture's loan guarantee program. The LCOE framework is also used to describe any avoided social and environmental costs, social costs of carbon, avoided criteria air pollutant emissions, and reduced utility regulatory compliance costs.

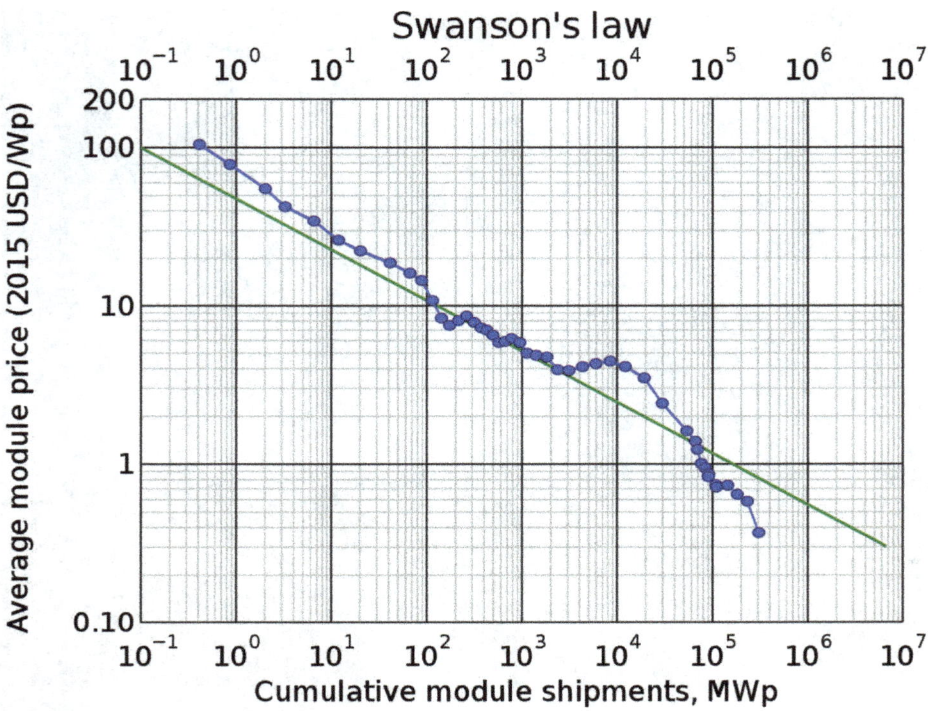

Fig. 7.3 Swanson's effect suggests that every doubling of production results in a 20% decline in cost

7

LCOE is a helpful way to compare electricity costs from different technologies. Not all electricity is equal; the price of electricity generated overnight in comparison to daytime generation from solar power may be incommensurate, unless you also include some kind of energy storage that can capture sunlight and turn it into power at night.

Assignment 7.1 Solar Energy Analysis Using National Renewable Energy Laboratory's System Advisor Model

Part 1: *Role of Financing terms on LCOE*

1. Read this: ▶ https://sam.nrel.gov/
2. Open SAM and run the solar wizard.
3. Select PV.
4. Select Residential sector.
5. Select a location

6. Select a system size: __________ kW_{dc}
7. Select a cost to install the project: _________ US$/W
8. Select payment type.
9. Under tax incentives choose the policy and amount of savings on installed cost.
10. Enter your approximate cost of electricity: ____________ US$/kWh

Annual output: ____________ kWh

Capacity factor: ____________ %

LCOE: ____________ c/kWh

Click start over and repeat 1–10, but instead of cash, you need a loan. Use default or change terms.

Down payment _________ %

Loan rate _________ %/year

Loan term _________ year

Annual output: ____________ kWh

Capacity factor: ____________ %

LCOE: ____________ c/kWh

Repeat 1–10, but instead of cash, your loan will be part of your mortgage. Use similar loan terms.

Down payment _________ %

Loan rate _________ %/year

Loan term _________ year

Annual output: ____________ kWh

Capacity factor: ____________ %

LCOE: ____________ c/kWh

Repeat 1–10. Use cash, but assume the utility offers a feed-in-tariff of US$0.12/kWh.

Annual output: ____________ kWh

Capacity factor: ____________ %

LCOE: ____________ c/kWh

11. Select PV.

Annual output: ____________ kWh

Capacity factor: ____________ %

LCOE: ____________ $/kWh

_________ %

Loan rate _________ %/year

Loan term _________ year

Annual output: ____________ kWh

Capacity factor: ____________ %

LCOE: ____________

Reflect on the observations from this analysis. What are the key variables that shape LCOE outputs? How do public policies impact these?

7.3 Power Density

Power density is one way to describe how much power can be obtained from an energy source per unit land or space. Technically, it represents the power flux per unit of horizontal surface. It is also power measured as the release per unit volume/mass; this is common where, for example, batteries used in vehicles are measured. Some units used to express power density include some amount of energy (joule, calorie, btu, kWh) per unit weight (gram, pound, ton) or volume (cubic centimeter, decimeter, cubic meter). The concept is helpful to provide a sense of whether specific energy resources can be matched to power some area, facility, or place. Just as some energy resources are more power dense than others—nuclear, fossil fuels are the most power dense—especially compared to renewables. Of the renewables, solar power is the most energy dense, though depending on the specific project and how its measured, sometimes its hydropower. Wind and geothermal power have lower densities (though geothermal has a relatively small surface footprint), while biofuels have the least density of power and take up the most land per unit energy.

Assignment 7.2 Power Density and Electricity Spatial Footprints

1. Calculate the power density of a poplar tree plantation used for electricity generation. Assume a yield of 20 tons per hectare per year and the energy density is 18 GJ/ton. Report this in W/m^2.
2. Assume a 1 GW coal-fired power plant is converted to a wood-fired one that can take the poplar grown on the above plantation. Assume the capacity factor is 85% and the retrofit allows for a conversion efficiency of 48%. How many square meters would you need to harvest to feed the plant annually?
3. If the surface of the Earth absorbs 240 W/m^2 and Californians use 10 kW per capita, what is the area needed to provide solar power to 40 million Californians?
4. Consider an average family of four living in the northern half of Europe. If the family harvested wood from an apple orchard and burned it, could the family harvest enough energy to heat its home? What conclusions can you draw? Use the following assumptions:
   - The family uses about 125 million BTU of heat energy annually to heat its home
   - The orchard yields about 3 dry pounds per person per day (assume 365 days/year)
   - This cured wood has heat content of 8,000 Btu per pound
5. The US has a lot of federally owned land. Unfortunately, forest fires destroy large areas in the Western US every summer. In 2002, about 1 million acres of standing timber in national forests were consumed by fire. Some consideration is being given to improved management practices that could produce electrical power from residual forest thinning for fire management. (1) Estimate the lost energy content of burned

US forests during 2002. (2) Assuming that the US average electricity demand load is about 300,000 MW, how much forested land would be needed to produce all of the country's power? Is this a sustainable alternative? A few facts to consider:

- The total forested area on US federal lands in the lower 48 states is about 600 million acres with a standing stock density of about 100 dry metric tons of wood per acre.
- Woody plants and trees capture solar energy via photosynthesis at an average rate of about 0.8 W/m^2, which corresponds to producing about 10 dry tons of biomass per acre annually. This wood has an average heating value of 8,000 BTU/lb. Note that 1 acre = 43,560 ft^2 = 0.405 hectare = 4,047 m^2.
- A representative heat-to-work conversion efficiency of a biomass-fired electric power plant is about 45%

Energy density is similarly a useful metric to look at various technologies with different capacity factors. Common metrics include kWh/m^2, or sometimes with solar energy potential, it is reported as energy per unit energy installed (kWh/kW). The best way to think about this is as a representation of how much energy can be delivered per 1,000 watts of photovoltaics installed in different locations with different solar energy resources (sunny vs. cloudy places).

7.4 Designing Electricity Systems for Flexibility

For future electricity systems, flexibility in both energy generation and demand will be critical. This is because of the intermittency associated with renewables. A cloud or rain system may move in and shade a solar panel, lowering output temporarily. Wind speed can fall sharply, and since its power is raised to the cube, this means an even more rapid decline in power. The duck curve is a representation of California's total electric load, showing how much of the demand is fulfilled by utilities in total (blue line) and after solar and wind power is excluded (red line). Over the course of a 24-hour day, the graph illustrates the imbalance between peak demand that occurs between hours 18 and 21, and renewable energy production. However, it is critical to understand why the duck curve is important and concerning for the current and future trends of California's electricity grid. Referring back to the red line representing net load, there is a very steep upward slope from hours 16 to 20. This best illustrates the situation that energy grid operators are facing. Because much of the demand from the grid is satisfied by solar and wind energy, other conventional peaker plants aren't needed to help provide electricity during the sunlight hours. However, during the evening hours, solar energy drops off and so grid operators are forced to "ramp" up energy production after sunset (Fig. 7.4).

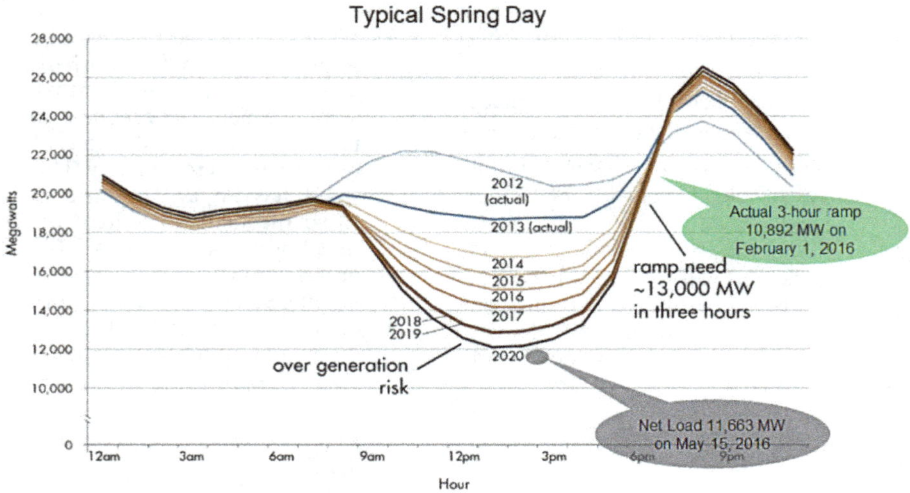

Fig. 7.4 California's duck curve shows the need for a significant power ramp up

For instance, solar energy, while providing much of the power to the grid, will not consistently produce the same results since there are days where it is either cloudy or there is no sunlight at all. The same can be said for other renewable energies such as wind, since there are times where there may be a strong breeze, and, at other times, the weather is stagnant. The varying factors on a daily basis make it difficult for operators to gauge how much energy is required from them in order to meet electrical demands.

Because the electrical demand isn't as easy to predict as it was in the past, grid operators have to "ramp" up the production in the sunset hours. However, having to ramp up the utilities to generate the power that is needed is quite expensive and actually generates more air pollution. Nuclear and coal power plants are slow and expensive to shut down. There are some natural gas power plants that are also relatively expensive to run, but they are much easier to ramp up and down as compared to the much older nuclear and coal power plants.

Metaphorically speaking, in order to resolve the issue of the sitting duck curve, the duck must learn how to fly by "implementing strategies to both flatten the load and to introduce supply resources that can deliver more output during the afternoon high load hours" (Lazar 2014). Jim Lazar's "Teaching the 'Duck' to Fly," are ten low carbon strategies that can be applied to address the issue of long-term sustainability of variable energy resources.

While each strategy involves audience participation of those who are directly or indirectly involved, the effect is much greater when several of the ten strategies are combined. They are (1) target energy efficiency to hours of the day when load ramps up, (2) orient solar panels to the west, (3) substitute solar thermal with a few hours of storage in place of some projected PV generation, (4) implement standards that allow the grid operator to manage electric water heating loads to shave peaks, (5) require new air-conditioners to include two hours of thermal storage, (6) retire inflexible generating plants with high off-peak must-run requirements, (7) concentrate utility demand charges into the "ramping hours" to enable price-induced changes in load, (8) deploy electrical energy storage in targeted locations, (9) implement aggressive demand-response programs, and (10) use interregional power transactions to take advantage of diversity in loads and resources.

While each strategy can result in outcomes towards energy sustainability, the combination of several strategies will have a greater effect than the sum of the parts. Out of all of the strategies listed, one of the most important ones to take note of is the substitution of energy generation with energy storage, ramping up to deliver significant amounts of power for several hours as and after the sun sets. A similar but less pronounced ramp down occurs in the morning as the sun rises toward the zenith, and other power plants must dial back power output.

7.5 Energy Storage and Integrated Resource Planning

The key challenges both to the duck curve and to the integration of intermittent energy resource are some way to store the energy generated, at times when the renewable energy resource supply is high, but the demand might be low. There are numerous kinds of different storage technologies, ranging from those for heating and cooling to those for electricity (McPherson et al. 2018). For electricity, the most common form of energy storage is pumped storage, which stores gravitational potential energy as water is pumped uphill during times of excess energy, to utilize the electricity later when there are peaks or voltage drops to attend to. Many pumped storage projects raise the same questions that are encountered with hydropower development (▪ Fig. 7.5).

Sodium-sulfur batteries have been used in electricity generation system in Japan since the 1980s. Flywheels are a storage technology that utilize kinetic energy as a wind-up toy stores energy. Compressed air technologies come in various forms, with some using natural gas, while others use different compression and expansion technologies.

Flow battery is another area of research and venture capital investment in the storage sector. Vanadium redux batteries use an electrolytic solution to hold the electric charge. While numerous sustained efforts continue to investigate this technology, very few have been able to compete with the low-cost batteries made from lithium ion (Li-ion).

As Li-ion batteries get cheaper, there will be increased synergies between electric vehicles (EVs) and stationary storage applications. Both rely on similar technologies and scalable configurations, even though they may have different equipment because of the particular applications used. Stationary storage batteries may even get a second life after EVs, as batteries are used in secondary markets that do not require the same performance and power delivery as seen with EVs. Stationary storage may be an ideal application for this, and repurposed devices may have significant cost savings (Ahmadi et al. 2014).

Fig. 7.5 There are proposals to turn Hoover Dam—a twentieth-century powerhouse for the American West—into a pumped storage project

As batteries become cheaper they will reward other renewables technologies as they essentially become batteries collecting energy from renewable sources at the same time they sell stored electricity to customers where it is needed on the grid, and in a way the grid prefers. Critics argue this will be extremely costly, as the cost of batteries will far exceed other energy resources on their own; but paring with renewables may be the game changing duo needed to power more of society with solar energy.

7.6 Smart Grids, the Internet of Things, and Artificial Intelligence

Progress toward cleaner and more efficient energy sources and practices will leverage technologies like the Internet of Things (IoT) and Artificial Intelligence (AI). Over the past decade, there has been increased investment in clean and renewable energies, as well as the emergence of a burgeoning IoT market driven by advances in AI. A popular application of IoT includes home automation innovations like Nest and Alexa. This technology has been extended into the enterprise market on a larger scale, using analytics to predict cooling and heating patterns to keep buildings at optimal temperatures, while utilizing the least amount of energy possible. While IoT devices generally utilize AI to enhance their operating capabilities, AI can independently play a significant role in promoting energy efficiency by becoming the brain, making decisions about operational parameters in future smart grids (Wolfe 2017).

Demand response (DR) will play an important role in future energy systems (Fallahi and Smith 2017). Load shedding and load shifting are two flexible DR applications, and they are essential for intermittent supplies to keep up with the energy demand (Müller and Möst 2018). DR can help reduce curtailment of power when there is overgeneration. Load shedding occurs when peak demand is reduced permanently. Load shifting occurs when power demand is shifted to times when demand is low or renewables are high.

DR programs fall into two categories: incentive-based programs and price-based programs. Incentive-based programs give consumers incentives from the distribution operator to reduce their consumption (Bajool et al. 2017). Price-based programs allow the distribution operator to change the electricity price through different tariffs and decreasing the system load.

Vehicle-to-grid (V2G) capabilities for EVs are a form of DR (López et al. 2015). In 2050, EVs could make up around 65% of the global transportation sector, or maybe a lot higher. As battery prices decline, more people will transition to EVs. More charging stations will connect more EVs to the grid.

Dispatchable, incentive-based programs and non-dispatchable or time-based programs are the two main categories of DR. For dispatchable DR programs, consumers allow system operators to control electric devices directly, like heating and ventilating air-conditioners (HVACs) during peak periods. These schemes include direct load control (DLC) and interruptible/curtailable service (I/C).

If intelligent charging schemes are adopted for car batteries, then the lifetime of the battery can be extended and will positively improve the grid as well. Smart grids rely on DR because by using information, communication, and control systems together, it can be better optimized. Electricity operators can implement strategies to deliver more output during hours of high load. Encouraging residents or industrial customers to change their load patterns is a key DR strategy.

DR programs include time of use (TOU), critical peak pricing (CPP), and real-time pricing (RTP), and these programs essentially offer a lower price during off-peak periods and higher price during high-peak periods. There are technical characteristics for the different applications that can be used for DR, such as night storage heater, warm water heating, and heat pumps, just to name a few (Muller & Most). For example, certain technologies might not last as long as other technologies for their demand shift; night storage heater and heat pumps have a 12-hour duration, air-conditioning has a 1-hour duration, and wood pulp production has a 2-hour duration.

The Federal Energy Regulatory Commission (FERC) is mandated to do an annual survey of DR and advanced metering in the US as per the Energy Policy Act of 2005. According to the American Council for an Energy-Efficient Economy (ACEEE), the peak demand savings average for DR programs is 10%, and this can be achieved if the utilities choose to use available resources to reduce energy consumptions (Nadel 2017). Bringing utilities and policymakers together is important to make DR programs more popular for consumers to buy into the programs, whether it is incentive-based or price-based.

IoT, coined in 1999, represents many classes of devices that are connected to a network, typically the Internet. These span the categories of wearables, cars, appliances, and home automation with the purpose of "making their user's lives more convenient" (Waltzman and Shen 2015). In addition to convenience, IoT serves to improve the safety of everyday life through autonomous vehicles which will reduce fatal car crashes and proactive sensor devices which can quickly alert monitoring agencies of threats, among others. IoT can be leveraged alongside artificial intelligence to analyze trends in data and optimize how devices are utilized to promote more sustainable and efficient practices. AI is a broad classification of technologies, which is described as intelligence demonstrated by machines. It is most commonly understood as computers that can make advanced decisions and can be compared to the cognitive natural intelligence, but many challenges remain. These are not just technological but also ethical questions.

Assignment 7.3 Research Paper on 2050 Sustainable Energy Strategies

You have been hired as an analyst-consultant to evaluate a technology, energy source, or energy sector that can help reduce the overall GHG emissions in California the state, region, province, or country by 80% by 2050. Describe the potential for your topic California and describe realistic deployment scenarios. The paper must describe some of the primary environmental and social impacts and the ways that impacts from deployment can be mitigated to ensure sustainable deployment. The impacts described should note how this strategy affects GHGs, air quality, water, toxicity, land use, and/or some other environmental metrics. The paper should also describe possible political, policy, and/or behavioral responses to the topic. What policies would you prescribe to deploy your strategy more broadly? What are the political interests involved and how will you mobilize the public to support your strategy? If the technologies are controversial (nuclear energy, clean coal, large-scale solar energy, wind farms), the most controversial aspects must be addressed to some degree. Consider cost: how much would this technology cost to be deployed and where will the investment come from? How could you attract investment or raise revenues to pay for things? At some point in the paper, you should define what you mean by sustainable energy too! **Note to instructors:** See www.dustinmulvaney.com/set for example research assignments.

References

Ahmadi, L., Fowler, M., Young, S. B., Fraser, R. A., Gaffney, B., & Walker, S. B. (2014). Energy efficiency of Li-ion battery packs re-used in stationary power applications. *Sustainable Energy Technologies and Assessments, 8*, 9–17.

Bajool, R., Shafie-khah, M., Gazafroudi, A. S., & Catalão, J. P. (2017). Mitigation of active and reactive demand response mismatches through reactive power control considering static load modeling in distribution grids. In *Control Technology and Applications (CCTA), 2017 IEEE Conference on* (pp. 1637–1642). IEEE.

Branker, K., Jeswiet, J., & Kim, I. Y. (2011). Greenhouse gases emitted in manufacturing a product—a new economic model. CIRP annals, 60(1), 53–56.

Fallahi, Z., & Smith, A. D. (2017). Economic and emission-saving benefits of utilizing demand response and distributed renewables in microgrids. *The Electricity Journal*, 30 (9), 42–49.

Howarth, R.B., (2005). Against High Discount Rates. In *Advances in the Economics of Environmental Resources*. Advances in the Economics of Environmental Resources. Bingley: Emerald (MCB UP), pp. 99–120.

Lazar, J. (2014). Teaching the duck to fly. Regulatory assistance project.

López, M. A., De La Torre, S., Martín, S., & Aguado, J. A. (2015). Demand-side management in smart grid operation considering electric vehicles load shifting and vehicle-to-grid support. *International Journal of Electrical Power & Energy Systems, 64*, 689–698.

McPherson, M., Johnson, N., & Strubegger, M. (2018). The role of electricity storage and hydrogen technologies in enabling global low-carbon energy transitions. *Applied Energy, 216*, 649–661.

Müller, T., & Möst, D. (2018). Demand response potential: Available when needed? *Energy Policy, 115*, 181–198.

NREL (National Renewable Energy Lab). (2018). pvwatts.nrel.gov

Smith, K. R., & Haigler, E. (2008). Co-benefits of climate mitigation and health protection in energy systems: Scoping methods. *Annual Review of Public Health, 29*(1), 11–25.

Timilsina, G. R., Kurdgelashvili, L., & Narbel, P. A. (2012). Solar energy: Markets, economics and policies. Renewable and sustainable energy reviews, 16(1), 449–465.

Waltzman, H. W., & Shen, L. (2015). The internet of things. *Intellectual Property & Technology Law Journal, 27*(7), 19–21.

Wang, M. (2005). Energy and greenhouse gas impacts of fuel ethanol. *Argonne National Lab, 23*.

Low-Carbon Mobility

Contents

D. Mulvaney, *Sustainable Energy Transitions*,
https://doi.org/10.1007/978-3-030-48912-0_8

Learning Goals

This chapter looks to get readers thinking about the future of how to get around and:

- describe how low-carbon vehicles work, their benefits, and areas of potential improvement;
- complete higher order calculations of carbon intensity;
- describe liquid and gaseous fuels from renewable energy sources; and
- appreciate the scale and magnitude of energy transformations for transportation and think about other ways to get us and our things around.

Overview

A tremendous amount of energy is used for getting us and our stuff around. Humans have always been a mobile species, always on the go. But the advent of cheap energy coupled with our new transportation technologies let humans move farther, faster, and more often. As we described earlier, cheap modes of transportation have resulted in deleterious environmental impacts from energy extraction, pollution, and infrastructures built to move people and stuff around. Transportation and its energies also have structured social relations in unequal ways. In the US, cheap energy helped facilitate suburban sprawl, leading to urban pollution when all these suburban cars were used each day in the city. On the flip side, some designs of transportation systems deepen racial inequalities such as with the infamous case of Robert Moses intentionally designing highways, specifically to prevent buses from New York City from traveling to the suburban and rural parts adjacent to the city, including many beaches to escape the heat of the city. This made it difficult for Blacks, other people of color, or other less well-off communities in the city to reach the beaches on Long Island to cool off during summertime.

This chapter focuses on the use of energy and transportation system, reviewing sources of energy used for transportation, and describes different ways of organizing infrastructure for future transportation systems. Electric vehicles (EVs), of course, play a key role. Of all the greenhouse gas (GHG) emissions produced from energy sources in the US, nearly one-third come from transportation with over 90% of GHGs from petroleum burned in gasoline and diesel-powered engines. Over 1.2 billion combustion engine vehicles on the road, in 2020, will need to be retired and replaced with more efficient and lower-carbon transit solutions. As of 2018, there are over 3 million electric vehicles (EVs) on the road, which is a good start but means there are still 1.2 billion cars to phase through. India and China will be major players in the low-carbon transportation revolution, with announcements already to prohibit the sale of cars using internal combustion engines (ICEs) by 2040. Professor Dan Sperling (2018) of the University of California, Davis Transportation Institute, says the future of mobility will be shaped by revolutions in automation, electrification, and decarbonization. The changes from the 2019–2020 coronavirus pandemic may reorient spaces to make more room for people.

8.1 Transportation in 2020 Is Powered Mostly by Petroleum

Why is transportation powered by petroleum? One important attribute of petroleum is that it is easy to store and transport. The fuels from petroleum—diesel, gasoline (known as petrol outside of the US and Canada), kerosene, jet fuel, and liquefied petroleum gas—have very high energy densities. Even with today's best batteries, a gasoline tank the same size of a given battery will contain several times more energy.

Definition

The future of mobility will be driven by new technologies that will facilitate (1) *automation* of driving and transport systems through advances in artificial intelligence; (2) *electrification* of automobility, which will make the translation of energy into wheel motion more efficient; and (3) *decarbonization* of transport by using more people-powered transport (walking and biking) and less carbon-intensive fuels, such as solar-powered electricity for EVs, hydrogen generated from clean energy, and biofuels.

There are social reasons that societies and regions become committed to certain fuel arrangements as well. Some countries use more petroleum than others, and the US is the leading consumer of gasoline and diesel fuel. Vaclav Smil (2011) argues that the US excessively consumes oil products compared to Europe because of (1) its inefficient fuel fleet, (2) virtual absence of more efficient diesel cars, and (3) complete absence of high-velocity trains.

There are many different proposed ways suggested to get people around with lower GHG emissions. The attractive areas in this space to investors and entrepreneurs include EVs, batteries, devices that utilize the Internet of Things (IoT), and autonomous vehicles (AVs). But the mainstay everyday practices of people—biking, walking, and telecommuting—may have a bigger impact. Infrastructure critically plays a role because if parts of Earth humans inhabit become more walkable and bikeable, many of the technological solutions could take care of the rest of the major remaining impacts. Look around, you are seeing new arrangements today to manage the pandemic. These are very low-tech solutions.

The paths to decreasing impacts from transportation focus on three areas. First, driving less, with more biking, walking, and telecommuting (or avoided travel). This requires significant changes to lifestyles and patterns of travel, which often reflect decisions about infrastructure or zoning. Second, it will be critical to replace combustion engines with electric vehicles to make transportation more efficient. Finally, the sources of energy that power vehicles must be decarbonized. This means producing electricity or fuels with low-carbon sources of power. These three tasks will reduce the bulk of the GHG emissions associated with mobility.

These considerations are very technologically centered, when in reality the complex social entanglements that make up extractive activities help us understand how these social forms are reproduced. To ensure a just transition requires building effective institutions that can help issues that can offer relief from the violence and human rights violations associated with the oil complex. Some corporate social responsibility efforts may be helping in this space, but the political dynamics are complex, and many activities are non-transparent.

8.2 Electric Vehicles

The shift toward EVs away from combustion vehicles will have significant public health benefits since more than 90% of criteria air pollutants are from combustion. Falling costs for energy storage are increasing the odds that most automobiles sold by 2050 will be EVs and not some other low-carbon-fueled sources like biofuels or hydrogen. In 2016, the sale of EV passed one million. A half year later, another million, and only a few months later, three million. This growth is poised to continue as places with large populations like China and India are banning combustion engines and sending signals to increase the demand for EVs.

EVs have lower operating costs than internal combustion engine (ICE) vehicles because they use electricity more efficiently. EVs have the added benefit of lower maintenance costs as they require fewer trips to the mechanic. More than likely, the emissions will get cleaner over time as more electric power comes from renewable, noncombustion sources. ICE engines get less efficient over their lifetime, owing to wear and tear, and become more polluting over time.

> The full life-cycle economic cost of a vehicle is its cost plus the cost of fueling it over some determined time.

The recent boom in EVs is driven by significant cost reductions and performance improvements in lithium-ion (Li-ion) batteries. Technical advances have led to several doublings of energy capacity of batteries and stark cost declines. Li-ion batteries are also widely used in various electronics, so EVs benefit from all the attention from different industries. The capacity to hold charge over many cycles and their position as number three on the period table (very light!) make them ideal metals to base batteries on. It is important to note that these batteries do not contain very much lithium as a percentage of the total battery. A list of common battery chemistries classified as Li-ion are lithium manganese oxide used by Tesla, Nissan, and General Motors; lithium nickel cobalt oxide used by Toyota; and lithium iron phosphate used by BYD.

Electrifying transportation will not be without its challenges. Finding sites to charge in urban areas can present challenges in the densest parts of cities. They may not have driveways, and wires crossing streets pose tripping hazards and would likely raise some aesthetic resistance. Although many would argue this is where there should be no cars at all! Even homes with driveways may not have an electrical outlet with 20 feet or so of the parking space for the vehicle. The odds of having an electrical outlet that can utilize higher amps and volts are even lower. This is why building codes are so important.

Battery density challenges mean that the range of EVs can be limited compared to gasoline. Range anxiety is an important issue that faces EV adoption. How far apart should be the charging stations? Can EV charging infrastructure be developed in rural areas? Rural drivers often have more long-distance travel patterns since most vehicle trips are short, and batteries can be topped off with every return home or where charging infrastructure may be present. The good news is that everyone has an electric outlet at home, and future homes will be built better and may include faster charging equipment.

The time it takes to charge an EV battery depends on the size of the battery but also the "level" of charging. The level 1 charger used in a common household electrical outlet adds four or five miles of driving per hour of charge. Level 2 units can add 10–15 miles of driving per hour or more. New level 3 stations add 80–100 miles of charge. The goal is to approximate stoppage times for shopping or a short meal, so make for more suitable refueling options. Level 4 chargers are supposed to rival stopping at a gasoline station.

Example 8.1 Estimate the Equivalent Power Flow of Pumping Gasoline

Imagine pulling up to the pump at a gasoline station and pumping ten gallons of gasoline into your tank in three minutes. What would be the equivalent power exchange if instead of gasoline we were using electricity? First, convert the energy of ten gallons of gasoline into MWh using dimensional analysis. Next, convert three minutes to hours. Since there are 60 minutes in an hour, 3 minutes = 3/60th or 1/20th of an hour. A total of 3 minutes is 0.05 hours. Finally, divide MWh by hours to get MW or power flow.

0.35 MWh divided by 0.05 h = 7 MW

This means the act of pumping gasoline into an automobile is the equivalent of pulling 7 MW of power. Compare this to the typical charging power of a level 1 charger, which is a little over 7000 W or 0.007 MW.

Battery swapping is another way that EV companies are thinking about solving this challenge of slow charge times. In this model, a driver would pull up to a refueling station, and the battery would be owned by a service that would swap the battery out with a charged one.

Other options including green hydrogen vehicles and biofuels may play a role as well, especially for heavy-duty transportation. These are discussed in more detail later in this chapter. Using tools already introduced and trade-offs between different energy sources and their impact on the environment can be evaluated to optimize future transportation systems. For a comprehensive assessment of these trade-offs, researchers have developed a framework called well-to-wheel analysis to compare different vehicle and fuel combinations.

8.3 Well-to-Wheel Analysis

Life-cycle assessment (LCA) is common to compare emissions from vehicles and fuels in the transportation sector. Well-to-wheel (WTW) analysis is a specific LCA framework to understanding the environmental impacts of fuel and vehicle combinations. The general idea is to capture all the inputs and emissions along the stages of production for both the fuel and vehicle combination. The findings from this type of research show that the fuel in an automobile, in general, yields the highest life-cycle impacts. WTW analysis framework can lend itself to comparing emissions from different fuel and vehicle combinations. To compare different fuels, researchers and regulatory agencies have conducted well-to-tank (WTT) analysis.

Assignment 8.1 Well-to-Wheel Analysis

Part 1: Compare the WTW emissions of the following 16 future low-carbon vehicle-fuel mixes to conventional vehicles on conventional gasoline today.

Part 2: Plot your results in Excel or Google Drive as a bar graph with *each vehicle and fuel mix* labeled on the x-axis and the *WTW annual GHG emissions* from each of these vehicles in the y-axis if they are driven 12,000 miles per year.

Hybrid-Fuel Mixes—these vehicles only use liquid fuels. Each vehicle should be listed in row in the spreadsheet.

1. Conventional ICE vehicle using the gasoline/ethanol blend available in 2020: 10% ethanol (E10) produced in the Midwest US; 90% gasoline.
2. Hybrid ICE vehicle using the gasoline/ethanol blend available in 2020: 10% ethanol (E10) produced in the Midwest US; 90% gasoline.
3. Hybrid Flex Fuel vehicle using 85% ethanol (E85) produced in the Midwest US; 15% gasoline.
4. Conventional ICE vehicle using the gasoline/ethanol blend of 90% gasoline and 10% cellulosic ethanol from poplar trees.
5. Hybrid ICE vehicle using the gasoline/ethanol blend of 90% gasoline and 10% cellulosic ethanol from poplar trees.
6. Hybrid Flex Fuel vehicle using 85% ethanol (E85) obtained from cellulosic ethanol extracted from poplar trees; 15% gasoline.

Use the following column headings in Excel:

Energy fuel-fuel density;

Energy % density carbon intensity;

Energy density fuel;

Vehicle annual miles; Fuel economy; EtOH mix %; $MTCO_2e$; Miles; gCO_2e/Mile (miles/gal); MJ/mile; MJ/gal; gCO_2/MJ; Volume MJ/gal; gCO_2/MJ gasoline; gCO_2/MJ EtOH; EtOH MJ/gal;

Energy density of gasoline = 132 MJ/gallon; Energy density of ethanol = 89 MJ/gallon; Carbon intensity cellulosic trees = 20.40 gCO2e/MJ; Conventional ICE fuel economy = 35 miles/gallon; Hybrid ICE fuel economy = 54 miles/gallon;

Drive cycle/annual miles = 12,000 miles/year.

EV-Fuel Mixes—these vehicles only take electricity. Add each as a row under a new set up headings for vehicles that take these fuels.

EV electricity on-peak EV electricity off-peak; EV PV charging; H_2 fuel cell vehicle (FCV) wind electrolysis; H_2 FCV electricity off-peak electrolysis; H_2 FCV natural gas/methane steam reforming (renewable feedstock); H_2 FCV natural gas/methane (natural gas feedstock)

Use the following row headings in Excel:

Fuel vehicle;

Annual $MTCO_2e$;

Annual miles;

gCO_2e/mile;

Fuel economy (miles/kWh);

Electricity CI gCO_2/kWh;

CI EtOH;

EtOH gasoline %;

CI fuel.

Some data to fill out the fuel/energy data: Carbon intensity (CI) of Pacific Gas & Electric (PG&E) electricity on peak = 428 gCO_2e/kWh;

CI of PG&E electricity off peak = 187 gCO_2e/kWh;

CI of PV charging = 15 gCO_2e/kWh;

CI H_2 FCV wind electrolysis = 10 gCO_2e/kWh;

CI H_2 FCV CA electricity off-peak electrolysis = 187 gCO_2e/kWh;

CI H_2 FCV nat gas (renewable feedstock) = 273 gCO_2e/kWh;

CI H_2 FCV nat gas (nat gas feedstock) = 511 gCO_2e/kWh;

EV fuel economy city = 4.0 miles/kWh;

H_2 FCV fuel economy city = 4.0 miles/kWh.

PHEV-Fuel mixes

Evaluate the following plug-in electric vehicles (PHEVs):

PHEV gasoline 2020;

Electricity on peak;

PHEV gasoline 2020;

Electricity off-peak PHEV; Cellulosic gasoline 2020; electricity on peak PHEV; Cellulosic gasoline 2020; electricity off peak

Use the following row headings in Excel: Then (1) paste red cells in columns F–H from the header in the EV section and (2) paste blue cells in columns J–T from the hybrid section. You will assume that gas supplies 25% of the overall energy; 75% of the time the PHEV is running in EV mode. You will calculate the annual $MYCO_2e$ from the EV mode operation (75% of the 12,000 miles) and the $MTCO_2e$ from gasoline mode, then add them to compare with the other results.

Annual miles;

$MTCO_2e$ Annual from EV;

Fuel vehicle $MTCO_2e$ mode;

Annual $MTCO_2e$ from gasoline mode;

Annual miles;

% on EV mode.

Part 3: Now plot your results. Add a second sheet to your Excel file. Copy and "paste special" (set to "values") the results from the calculator into one master list on that second sheet. You should have two columns. One listing all 16 fuel vehicles, and one listing the annual $MTCO_2e$. Go to CHARTS. Click BAR and select CLUSTERED BAR. Give your graph a more appropriate title and you're done.

Part 4: Everyone in class will join a collectively shared Google Sheets or equivalent. This sheet will be the master form for all of our sustainable energy strategy calculations. In this sheet, begin to populate the total list of vehicles in the 2050 fleet. One idea might be to have a separate line for each class of transportation. This would allow you to require cars get more fuel efficient over time. Treat the annual miles traveled as the total vehicle miles traveled by each vehicle type. If you prefer, you can begin to make these vehicle types very specific. To make it for all of place you are aiming to assess California, we'll have to add in freight and rail transportation and add in new categories of transportation types, light trucks, heavy-duty trucks, public transportation, and so on. Let's build this with several hundred cars if possible. Here are some other odds and ends for now.

- Insert a column that reports the amount of fuel in gallons, mass of H_2, and kWhs used. This will be helpful later because for some fuels, you will have to be able to see how much can be produced in the state.
- Add another column with the heading "Water Use" because for each of these fuels, we will need to supply water. This increased water use will be added to the California water unless it displaces some agricultural production.
- Add a column for rare earth elements or other rare inputs. These can be sourced from the places you specify, under the preferred production standards. Also add fertilizer, land, and any other limiting factors.
- Add a column titled "MT electric batteries" (MT = metric tons) as each car type will have different battery amounts. Battery manufacturing will be a new major industry in California. Imagine that this California sustainable energy strategy is going to invite investors like Elon Musk to build giga-factories but also Panasonic, Toyota, and GM. Reno, Nevada, is close to numerous placer mine claims in the state of Nevada. In the future, it might become possible to source lithium for Tesla's giga-factory from these sites of extraction. Either way each team will have to estimate how many batteries go to all sectors and estimate the life-cycle energy required to manufacture them all.

8.4 Hydrogen Fuel Cells

Hydrogen is the most widely occurring element in the universe, constituting about 75%. Astrophysicists tell us that all hydrogen has been around since the universe was a few thousand years old. On Earth, there is plenty of hydrogen bound to other chemicals, but hydrogen gas is pretty rare. A key attribute that makes hydrogen a good fuel option is it is very energy dense, with a lot of energy packed in a small volume (or mass). Hydrogen has a higher energy density than gasoline. But hydrogen is also very volatile and poses safety challenges because of the potential for it to explode during storage or transport. Nevertheless, the production of green hydrogen from renewable energy could make hydrogen fuel cell technologies a part of the automotive fleet of a decarbonized world. There are several energy transition analysts who have proposed and touted the benefits of a "hydrogen economy" (Rifkin 2003). More recent conversations are talking about hydrogen as seasonal storage or in industry.

Definition

Green hydrogen commonly refers to hydrogen produced by electrolysis powered by renewable energy.

Hydrogen is an energy carrier, not a primary energy source. As a fuel, hydrogen can be used in several ways, including mixed with natural gas for combustion (e.g., for heating) or passed through a fuel cell for electric power generation. Hydrogen fuel cells take hydrogen and release energy, and they produce oxygen and water. They are functionally the opposite of electrolysis, where water is split into hydrogen and oxygen. Electrolysis uses electricity, and so its impacts are a result of electricity mix. The process of electrolysis will have an efficiency rate between 50% and 80%. New proposed designs for electrolysis processes utilize solid oxide electrolysis cells are aiming to achieve efficiencies of 90–95%.

Definition

Hydrogen fuel cell—a device that converts hydrogen and oxygen into water and produces electricity.

Hydrogen is energy intensive to harness from water, requiring more energy than is gained from the fuel. Most hydrogen today is from steam reforming of natural gas. It can also be produced by way of gasification of coal and from ethanol. Hydrogen can be taken from methane derived from decayed plant matter produced by photosynthesis. Methane from biogas at wastewater treatment facilities is primary biomass-derived material, although there may be some fossil fuel products in these waste streams, for example, plastics, used motor oil, industrial lubricants for machines, and so on.

Borrowing ideas from nature might offer ideas. University of Michigan researchers developed a device that uses artificial photosynthesis. The device is made of very similar components to photovoltaics and LEDs, such as silicon and gallium nitride. The gallium nitride transforms light (in the form of photons) into freely flowing electrons and holes (positively charged electron-vacancies). As photons hit the device, the electric field separates excited electrons from holes to produce hydrogen and oxygen. Natural photosynthesis has an efficiency of 0.6%, but this method they claim achieves 3% and in theory could be much higher. Since the device just uses sunlight and water, much like natural photosynthesis, it could be a pathway for large-scale production of clean hydrogen fuel.

A major disadvantage of hydrogen fuel cells for vehicles is inefficiency. After production, distribution, and storage, hydrogen energy is still only about a third of an electric vehicle's efficiency. However, despite the fact that a hydrogen fuel cell vehicle can charge significantly faster than a battery of an electric vehicle, 5 minutes opposed to up to 18 hours, there is a lack of infrastructure, such as hydrogen charging stations, that support the vehicle—electric car charging stations are much more accessible. With a lack of infrastructure, it will be difficult for consumers to be motivated to buy the cars *Nature*. Even if the cars are out in the market, it is very difficult to "establish a [loyal] customer base, increase production, [achieve economies of scale], and bring down costs" (Tollefson 2010). But this has definitely been a positive development for EVs, which is that people who buy them would buy them again, and most would never buy a gasoline car again. So long as the EV revolution is led by safe, reliable, and hopefully soon affordable cars, they seem destined to have increasing market share from here to 2050.

The distribution of a future automobility of battery-powered electric cars, hydrogen fuel cell-powered cars, and hybrids will be based on the resource size, availability, and economic potential of each option, and it may be regionally specific to the available fuels and infrastructure. It is not just the total amount of energy (resource size) that needs to be available, but also the time when that energy would be available, including factors such as the time of day, seasonal availability observations), miles and duration of driving, and new ideas about how roads should be used, which might not see a future for passenger cars at all except for those who could better accessibility.

8.5 Ethanol

Sugar- and starch-based platforms for liquid biofuels make up the most significant portion of the biofuel supply. The leading crops for ethanol production include corn and sugarcane, largely from the US and Brazil, respectively. Almost one-third of the US corn supply went to ethanol production in 2016 (USDA 2017).

Definition

Energy Content of Selected Liquid Fuels (in kilo-British thermal units (kBtu))

Gasoline	125 k Btu/gallon
Ethanol	84 k Btu/gallon
Compressed natural gas	106 k Btu/gallon
Propane	91 k Btu/gallon

To utilize ethanol, we can use flex-fuel engines, an internal combustion engine able to take higher percentages of ethanol, but also liquid fuels with high proportions of gasoline. The nomenclature that relates these high percentages of gasoline versus ethanol are E10 (10% ethanol), E100 (all ethanol), E90 (90% ethanol), and so on.

Example 8.2 Energy in Ethanol

The Energy Policy Act of 2005 is a US law that requires the production of 36 billion gallons of ethanol (EtOH) by 2020. (1) How much energy does 36 billion gallons of ethanol contain (in megajoules (MJ))? (2) If US drivers use the same amount of energy to get around, how many gallons of gasoline would this displace each year?

$$(1) \quad 36{,}000{,}000{,}000\,\text{gallon EtOH} \times \frac{80.2\,\text{MJ}}{\text{gallon EtOH}} = 2{,}890{,}000{,}000{,}000\,\text{MJ}$$

$$(2) \quad 2{,}890{,}000{,}000{,}000\,\text{MJ} \times \frac{\text{gallon gasoline}}{121.3\,\text{MJ}} = 23{,}800{,}000{,}000\,\text{gallons gasoline}$$

Many studies of GHG emissions from ethanol suggested that the energy balance did not yield positive returns. In other words, it had an energy return on investment (EROI) less than 1. The best corn ethanol EROIs are in the range of 2 and 3. The worst-performing ethanol plants for decarbonization are those that use coal for the heat process and have large amounts of coal on the power grid.

If energy crops remain major sources of liquid fuels, they will compete with high-quality land for agriculture, creating trade-offs between food and fiber and fuel production, or other land uses. During the rapid scaling-up of corn ethanol after the ethanol mandate, there were so-called “tortilla riots” in Mexico, not because of a shortage of corn, but from rapidly rising prices (McMichael 2009). This particular situation may have been a historically uncommon situation as there were several factors driving the price of food. The backdrop was the financial crisis of 2007–2008 when many investors were moving cash out of financial products like mortgage-backed securities into commodities; people having hedge funds and futures speculators were looking for safer bets, even if the returns were more modest. Also, major processed food companies began to hoard any available food on the market to ensure they remained profitable. It was a perfect storm when the increased demand for ethanol is accounted for.

Example 8.3 Energy Density and Well-to-Wheel Greenhouse Gas per Mile

1. What is the energy density of E10 fuel?
   Assume: Energy density of gasoline = 120,000 Btu/gallon
   Energy density of ethanol = 80,000 Btu/gallon
   Calculate energy density of the combination of ethanol (10%) and gasoline (90%) called E10 = 116,000 Btu/Gal
2. (a) How much well-to-wheel (WTW) life-cycle energy per mile is used to move a compact hybrid car, and (b) what are the annual GHG emissions associated with this vehicle if driven 12,000 miles annually? Assume the vehicle is running on an E10 blend. You will have to calculate the carbon intensity of E10 fuel.

WTT efficiency of E10 production = 80% = Amount of initial energy converted to E10; Inverse of EROI.
Carbon intensity of conventional gasoline = 92.7 gCO_2e/MJ
Carbon intensity of average corn ethanol = 72.6 gCO_2e/MJ
Fuel economy = 54 miles/gallon
Drive cycle = 12,000 miles/year

$$(1)\quad \frac{\text{WTW Energy}}{\text{mi}} = \text{WTT x TTW} = \frac{\frac{116{,}000\,\text{Btu}}{\text{gallon}}}{0.80} \text{ x } \frac{\text{gallon}}{54\,\text{miles}} = 2{,}700\,\text{Btu / mile}$$

$$(2)\quad CO_2e = \frac{12{,}000\,\text{miles}}{\text{year}} \times \frac{10.5\,\text{kg}\,CO_2e}{\text{gallon}} \times \frac{\text{gallon}}{54\,\text{miles}} = \frac{2{,}300\,\text{kg}\,CO_2e}{\text{year}} \text{ or } 2.3\,\text{MTCO}_2\text{e}$$

Food versus fuel debates could instead ask, is it possible to produce food *and* fuel? One past example is the German alcohol economy during World War II. Farmers would deliver potatoes to energy producers, who would distill these foodstuff to alcohol, then the farmer would get one-third of the biomass pulp for pigs. This is sort of what happens even in the ethanol economy today in the US, where distillers' grains are sold to the livestock industry, and there are some studies that suggest this food is better for animal health. The food versus fuel tension points to a need to develop more next-generation biofuel sources such as algae and cellulosic ethanol that do not compete with food or prime farmland. There are a handful of candidates for cellulosic ethanol. Sugarcane is a relatively less land-use-intense biofuel system, using niche wetland-type areas but deserves critical scrutiny too; the corn-ethanol system is not a good use of agricultural resources considering global problems with uneven access to high-quality chemical fertilizers.

Water quantity and quality issues are other considerably large impacts from producing biofuels. Water footprint studies of biofuels find very large water needs for growing many energy crops with the evapotranspiration associated with photosynthesis. Freshwater use is clearly more important in regions that rely on fossil rainwater harvesting from irrigation, such as agricultural operations growing corn for ethanol on the Ogallala aquifer in the Midwest US rather than rain-fed corn or energy crop production.

Nitrogen pollution is another factor, as farmers typically overuse cheap fertilizer to ensure that yields are maximized. Also, as ethanol emerged as a marketable product, many farmers brought nitrogen fertilizer back to the soil instead of bringing nitrogen by rotating corn with soybean, the latter mimicking the way nature moves nitrogen from the atmosphere into the rhizosphere. Likewise, where pesticides are used, there are concerns about residues in the soil and waterways, and data showing that these chemicals do not readily break down.

In the case of corn ethanol in the Midwest US, the major ecological and water quality concern is where land is taken out of conservation easements and brought into corn production. Conservation easements allow farmers earn revenue by leaving native grasses or prairie in place as a buffer to absorb nitrogen and soil runoff, and as forage and habitat for insects and wildlife. When agricultural prices are high, these lands might be taken out of easement. Or, if crops are designed for marginal land, these too could put easement habitat at risk.

Candidates for second-generation ethanol crops include miscanthus (*Miscanthus sinensis*), giant reed grass (*Arundo donax*), elephant grass (*Pennisetum purpureum*), poplar trees (*Populus* sp.), and switchgrass (*Panicum virgatum*). Some of these crops require more research and development in plant breeding, which will likely play a major role in making these feedstocks economically viable. Some of these objectives include increasing biomass, decreasing fertilizer requirements, optimizing yields, introducing enzymes into the plant, modifying/removing lignin, and converting cell wall to sugars.

One final consideration in evaluating the environmental impacts of biofuels has to do with plant breeding. It turns out many of the same traits sought-after in biofuel crops are similar to those found in invasive species: drought tolerance, tolerance to salinity in water and soils, ability to grow in low-fertility soil where prime farmland is not required, high water-use efficiency to lower the water footprint, increased harvestable biomass to increase the productivity per acre, increased seed production where seeds provide the fuel (biodiesel, next section), fewer pests and less susceptibility to diseases so fewer pesticides are required, and high density for greater land-use efficiency.

Third generation biofuels include micro- and macro-algae. Algae biodiesel or biocrude can be produced from numerous feedstocks, including diatoms, green algae, blue-green algae, and golden algae. These biofuels, which can be ethanol, biodiesel, and even biobutanol, can also be grown in the ocean or wastewater. Many of these algae species contain up to 50% body weight of lipids—fatty cells that can be used for biodiesel. These feedstocks can use CO_2 emissions as an input from other production processes to enhance the feedstock production. Algae purportedly has a high-power density at 2,000–20,000 gallons per acre per year, though it has not been possible to produce commercially to date due to affordability and scale.

Producing ethanol from waste biomass streams could make sense where there is an abundance of biomass availability, where it might already be collected and concentrated, and where the waste flows are regular and in large volumes. Some communities have biomass available through yard clipping programs or wood for wildfire fuel reduction or invasive plant removal efforts could provide supplies of energy. Cattails grown at wastewater treatment plants could be harvested and fermented with very high conversion rates.

Today, ethanol remains a contested potential energy source. Some of this disdain is because it remains a product of combustion, which still leads to some kinds of air pollution. Other concerns are raised because some corn ethanol sources might not yield back the energy or GHGs invested to make them. For now, ethanol production is interlinked with many regulatory standards but also a growing number of certification and labeling schemes and will be part of the energy mix for the foreseeable future (Bailis and Baka 2011).

One final consideration is that ethanol is added to gasoline in order to replace methyl-tertiary butyl ether (MTBE), a fossil fuel additive mixed in to lower the air pollution caused by combustion. Adding even corn ethanol to gasoline instead of MTBE should lower GHGs emissions in a gallon of gasoline. From 2002 to 2004, the EPA conducted a multiple-year LCA of impacts from swapping MTBE and ethanol and found that the air pollution and water pollution impacts from ethanol were much lower than MTBE.

8.6 Biodiesel and Renewable Diesel

Biodiesel and renewable diesel are actually different liquid fuels. Both are produced through a thermochemical conversion process, so they are in what is called the thermochemical platform for bioenergy. Transesterification is the process to turn triglycerides from fats and oils in the presence of methanol into methyl esters (biodiesel) and glycerin. Biodiesel is made of fatty acid methyl esters (FAME), which can be used in most engines in small quantities, but it requires some engine modifications for blends with high FAME content. The key difference between FAME biodiesel and renewable diesel is the presence of oxygen. Hydro-treating is the process of making renewable diesel, which has the same chemical formula as diesel does. This process does require high temperatures, so it can be more energy intensive to produce. B100 means 100% biodiesel, whereas B10 is 10% biodiesel. Much like the nomenclature for ethanol, the number stands in for the percentage. Renewable diesel, because it has the same formula, does not similarly have a nomenclature.

Global renewable and biodiesel feedstocks and production are largely restricted to a handful of countries and regions, including across the European Union, the US, Brazil, and Malaysia. Common feedstocks include canola, soybean, palm, hempseed, vegetable, and corn oil. Soybeans are one of the most common biodiesel sources, but they yield relatively small amounts of crop per land. Directing all the US soy crop (and the US is a major producer) would satisfy only about 5% of US biodiesel demand if it were to replace diesel. US annual consumption of diesel is about 60 billion gallons per year (Energy Information Agency 2017). The entire continent of Africa used 2.5 billion gallons in 2016.

Palm oil from Indonesia and Malaysia (which is also destined to the food industry) is impacting habitat for the critically endangered orangutan, a resident of the forests that are converted to palm plantations. In the mid-2000s, growth in palm oil production was spurred by EU policy, and much of this habitat came under pressure for development. Greenpeace and other groups began campaigns, and soon a Roundtable on Sustainable Biofuels team emerged specifically to work on sustainability certification schemes for palm for biodiesel. One challenge is that more palm oil goes to the food industry, so growers have alternative markets to sell to.

A number of second-generation biodiesel crops are being explored, including jatropha, algae, and other exotic oilseeds. These crops promise to have lower input requirements, as they are not the fertilizer-hungry crops typically used for biofuels. Likewise, these crops should be more productive per unit area. In the case of algae, these can be grown offshore or indoors. Several key policies have driven much of the development of markets for biodiesel over the past decade, including a blenders credit for biodiesel or vegetable oil, a small producer tax credit. But beware of policy design, google "splash and dash" biodiesel loophole for an example pitfall.

Straight vegetable oil (SVO) is a fuel made from "off-the-shelf" cooking oils that can be directly dropped into fuel tanks. While functionally it is good to use up waste, and most likely the fuel has very low carbon, there are resource limitations to the availability of some of the oil produced, and the costs to produce these oils are relatively high. Waste vegetable oil (WVO) is made from used tallow or food greases. This resource also is limited in its availability, but it could be used in niche energy applications and markets, for example, on farm or in fleets of vehicles near food processing centers or restaurants. If all of the greases and animal fats in the US were converted to biodiesel, it would replace about 1.5 million gallons of diesel fuel, which is not much at all (National Renders Association 2000).

Assignment 8.2 Environmental and Ecological Justice in Supply Chains for Biofuels, Electric Vehicles, and Hydrogen Fuel Cells

Biofuels, electric vehicles (EVs), and hydrogen fuel cell vehicles will provide cleaner, lower carbon options for transportation California. Yet, there are serious critiques regarding the sustainability and environmental justice issues in the supply chain for several of these prime mover and fuel combinations. John Sheehan (2009) writes, we need to "avoid the never ending cycle of solving problems with new technology solutions from which new problems arise that need new technological solutions." For this assignment, each student should pick one of these supply chains to investigate: (1) cobalt for EVs, (2) lithium for EVs, (3) manganese for EVs, (4) hydrogen for fuel cells, (5) platinum for fuel cells, (6) sugarcane for ethanol, (7) corn for ethanol, (8) soybean oil for biodiesel, and (9) palm oil for biodiesel.

In this assignment, each student role-plays what they would do if they are hired as a sustainability advisor to a policy-maker tasked with developing criteria to evaluate the sustainability and social justice dimensions of transportation options. The criteria you propose for sustainability and social justice will be used to benchmark and select which kinds of policies can be pursued to maximize ecological and social benefits of sustainable energy strategies. How can we pursue the prime mover and fuel options mentioned earlier without reproducing the social and environmental impacts they currently cause?

Write a 1,000-word summary with five peer-reviewed research articles or high-quality white papers and reports that (1) provides an overview of what kinds of transportation technologies could be in use in 2050 and describes end uses where they could replace fossil fuel energy; (2) highlights the major socio-ecological impacts of the assigned batteries/fuel cells/energy sources and major efforts to improve, mitigate, or minimize these impacts; and (3) proposes a set of criteria, regulations, and/or best practices to ensure most sustainable and socially just biofuels are produced for the chosen region's California's sustainable energy strategy.

Address your letter to an elected official or a policymaker. Focus on what you deem to be the most important issues to screen for based on your understandings of the readings. Your description of impacts need not be exhaustive, but it should contain the major considerations and emphasize the ones most important. Describe qualitative criteria, certifications, and quantitative metrics (use tables if it is easier to present metrics this way) and make sure you consider people and the environment.

8.7 Low-Carbon Drop-in Fuels

8

An obscure decarbonization proposal comes from a group of scientists promoting the methanol economy, which is a liquid fuel known as "wood alcohol" that can be produced by oxidizing methane or reducing carbon dioxide. Depending on the source of the methane and CO_2, this could reduce GHGs from these fuels compared to conventional gasoline or natural gas. Dimethyl ether (DME) can also be derived from methanol and can be used in gas turbines, and as a substitute in blends with diesel, gasoline, and liquefied petroleum gas (LPG). So DME offers the possibility of a drop-in fuel with minor modifications. Carbon waste streams, preferably carbon sourced via photosynthesis, can be converted to cleaner burning methanol, which contains no sulfur. To make these fuels possible, special catalysts are needed to facilitate reactions that release the hydrogen from methanol. There is also experimental research ongoing with methanol fuel cells. Much like the case of hydrogen, most of the methanol supply today is derived from fossil carbon, although numerous efforts are underway to produce bio-feedstocks for various alcohols, such as biobutanol.

A Nobel Prize–winning chemist Dr. Shah argues that the methanol economy would be a relatively easy transition from an infrastructure perspective because it is relatively inexpensive to convert existing gasoline and diesel service stations to methanol ones. The carbon for the methanol would ultimately be coming from the atmosphere. Methanol as a liquid fuel also overcomes the power transfer problem at charging stations for electric vehicles, where there are two orders of magnitude difference in power transfer when moving liquid gasoline to the tank compared to the ability to move electricity to an EV battery. Currently, there are no meaningful pursuits of a methanol economy with the exception of the more obscure research, likely because the feedstock for methanol will be natural gas for the foreseeable future.

8.8 Vehicle-to-Grid Storage

Electric vehicle batteries that are capable of two-way electricity flow into and out of the power grid can be used in vehicle-to-grid storage (V2G) applications. V2G integration can contribute energy storage capacity and provide services to the electricity grid. By providing energy, V2G can help provide key resources at important times, as V2G systems can be used for reducing peak loads and backing up power. This added storage can integrate intermittent renewable electricity generation into the grid by using a buffer of renewable energy in the batteries when it is available. The V2G system can also improve the quality of energy, preventing critical voltage drops or frequency response services.

Any value created by V2G could be used to incentivize PEV ownership and practice. Some of the opportunities could further reduce emissions in the transportation sector. Some modeling efforts suggest that widespread V2G deployment could benefit PEV buyers, electricity ratepayers, and society more generally. California is testing V2G projects to see how it will help it reach its climate goals by 2050. That allows them to go from simply consuming energy to potentially becoming a fully functioning component of the smart grid. The two-way functionality of electric vehicle batteries makes it an ideal source for energy and storage. This will give electric vehicles power when needed and give grids power when needed. This can be a way to reduce power at peak times when the grid needs to acquire or shed load. By 2050, there could be a 39% increase in non-transportation load and vehicle miles traveled due to population growth compared to current 2015 levels. An 80% BEV penetration increased the electricity load by 32%. The net effect is an electricity load 183.5% of the current CA load (Forrest et al. 2016). In the long run, more electricity will be used by 2050. Integrating the vehicle-to-grid technologies can store electricity and allow it to be used when needed and vice versa. In 2018, BMW partnered with Pacific Gas and Electric (PG&E) to pilot smart chargers that will charge batteries during the day, when solar energy is abundant and sometimes is curtailed. By subjecting smaller amounts of renewable energy to round-trip efficiency losses, and thereby increasing the efficiency of renewable utilization, vehicle-to-grid energy storage can help achieve higher renewable utilization and quality levels and reduce GHG emissions compared to "stationary energy storage systems" (Tarroja et al. 2016).

8.9 Autonomous Vehicles

How far will autonomous driving push our civilization with regard to economic and environmental considerations? Companies such as Google, GM, Tesla, Apple, Uber, and Volvo had made significant financial commitments to this technology, and they believe it to be a critical component to the future of mobility. Combining these two technologies—EVs and AVs—together into a single, interdependent package with infrastructural and legislative changes will drastically increase driving efficiencies, reduce driving accidents, and, with the help of electric vehicles, reduce GHGs. One study suggested that taxis from a fleet of AVs would have significant GHG reductions (Greenblatt and Saxena 2015). But studies of AVs in practice have many different outcomes, depending on the operating conditions and its urban, rural, highway speeds, or stop-and-go (Mersky and Samaras 2016).

At a high level, the idea is to utilize autonomous driving technologies with electric vehicles to create ride-sharing "pods," if you will, that will allow people commuting together to ride together. High-speed travel could change drastically increasing the minimum driving speed or moving chains of large trucks or buses. Urban travel could be based on vehicles moving at slow-paced minimum speed. Both could drastically reduce the number of vehicles on the road and also the GHG emissions, while at the same time requiring great resources and fewer parking lots.

There are still some technological breakthroughs that are necessary before autonomous vehicles (AVs) can become widely available or put into use including very important ethical questions. AI has been improving rapidly and has propelled the ability of AVs to analyze and react to a dynamic driving environment to the point where minimal human interaction is necessary. However, the capabilities are currently not advanced enough to become a viable solution in today's basket of mobility solution except in very low-risk situations.

Scenarios have been simulated to study the impact AVs will have on driving habits and GHG emissions. In some imaginations, the ideal scenario in terms of energy efficiency envisions the entire personal automobile fleet will become automated, only consist of ride-sharing, and will largely incorporate vehicles that have low emissions such as hydrogen fuel cell and electric vehicles. This will result in energy savings stemming from traffic flow smoothing, vehicle-to-infrastructure and infrastructure-to-vehicle communication, collision avoidance, and vehicle powertrain resizing. It is estimated that the US light-duty vehicle fuel consumption will decrease significantly, given the vehicles on the road are more fuel efficient. While this seems to be a likely outcome indicated by the successes in the adoption of electric vehicles and ride-sharing platforms, it is much too early to conclusively determine that AVs will result in major reductions of GHGs or improved energy efficiency.

The most concerning challenge facing the adoption of IoT or AV technology energy efficiency is the threat of cyberattacks. By connecting more devices to the Internet, especially ones that govern critical infrastructure like the power grid, we expose ourselves to attacks from political assailants and people up to no good. Bad actors would have a chance to exploit the various energy systems within an individual home, business, vehicle, or power grid at large. The issue of cyber security is both a political and technical problem, which has to be addressed with laws and technical safeguards (Singer and Friedman, 2014, 165). On the technical side, the US Senate has stressed that "security should be built into devices at the outset and throughout a device's lifecycle," but not as an afterthought. Although it is impossible to protect from all threats at the outset, good design practices can significantly reduce the likelihood of crippling attacks (Singer and Friedman, 2014, 55).

AVs will be connected by smart hubs spaced at some interval to ensure proper and adequate coverage to assure that each vehicle is communicating and is still on the network. These smart hubs act as the central device and hand off to the next hub is seamless from one to the next. As reliability and effectiveness of public transit via autonomous ride sharing increases, there will be more opportunities to smart transit design that can incorporate AVs. As the public engages with more autonomous driving, there could be much lower need for parking lots and structures.

As AVs have the potential to change vehicle design and use patterns, there is much potential for emissions reductions. But some suggest that AVs could increase overall emissions due to a rebound effect. The EIA reports that light-duty vehicle miles travelled could increase with AVs, leading to overall increases in emissions (Energy Information Agency 2017). A useful guide for policymakers has been developed by some of the leading experts on these topics (Anderson et al. 2014).

8.10 Public Transportation

The US constructed its first rapid transit system, a subway line, in 1895 in Boston which was 1.5 miles long and used trolley streetcars. Then, on October 27, 1904, New York City opened its own rapid transit system of what became one of the largest systems in the world. Los Angeles had an extensive public transit network that was dissembled.

There are even public health benefits documented with the use of public transportation. Transit users are more active, safer, and less stressed. The US Center of Disease Control (CDC) recommends about 22 minutes of physical activity and the average individual who uses transportation does 19 minutes from walking to stops and final destinations. Moreover, transits promote individuals in communities to interact and have access to social or recreational activities or participate in events that were not accessible initially. In addition, approximately 40,000 deaths happen annually from car accidents per year; however, traveling on a transit is safer and areas with high public transit use have security. The coronavirus pandemic is posing challenges for public transportation, and it has led to steep declines in revenues from travelers.

The kinds of technologies communities are pursuing to decarbonize transportation include electric buses. In the US, many diesel buses have already been replaced by natural gas, but much more opportunities exist for electric buses. Some cities like San Francisco have used electric buses for many decades. But these are the kinds of buses that require overhead wires. These impose a very visual blight on communities. Electric buses with a battery onboard or capable of wireless induction charging offer an opportunity to replace combustion engine buses with emissions-free electricity.

According to the American Public Transportation Association, every dollar invested into public transportation generated approximately US$4 in economic returns. These returns can be invested in communities to create jobs, increase business sales, and increase residential property values for homes near to the transportation systems, and for every US$1 billion in investments, 50,000 jobs are created. Public transport agencies employ tens of millions of people worldwide. The organization also reports that a person can save over US$10,000 per year on transportation costs by using transit. This is because a person only pays for when they are on the ride and do not have to worry about expenses such as maintenance, repairs, parking, or ongoing improvement.

Things are looking up for electrifying public transportation. Today, there are increasing orders for electric buses from cities and regional transportation agencies. It is expected that by 2020, one-third of all new buses will be electric. Major rails are on the move from diesel to electric in some places. The Caltrain route from San Francisco to Gilroy, California, is expected to be electrified in the near future.

8.11 Urban Planning for Walking and Biking

Perhaps the most efficient way to improve the sustainability of mobility in urban areas is to make them more bikeable and walkable. This will take substantial time as urban planning efforts can be slow and incremental, particularly in large cities. These again are areas where there is significant evidence to demonstrate public health benefits from walking and biking because of the associated cardiovascular exercise. It would be difficult to find a city that has done this with great success, but, increasingly, there are projects that have become exemplar, especially in response to coronavirus public health measures. These designs usually keep pedestrians, cyclists, and vehicles separated in some way. Ultimately, making streets safer will result in more people walking and biking in their communities.

Bikeshare and scooter programs are taking off in some parts of the US as global positioning system (GPS) and mobile currencies allow new modes of payment and tracking. These trips may be displacing traffic rides and taxis. It may allow someone to travel by foot one way and take a ride home in some circumstances. As new business model opportunities have emerged for these bikes and scooters, their rapid deployment in some places has drawn a bit of public ire because of clutter. Most onlookers seem to tend toward endorsing this new mode of transportation as a final mile solution.

Non-automobile transit-oriented communities are another opportunity for urban planners to design communities that maximized access to public transportation. If people have safe, efficient, and effective means of public transport or biking and walking opportunities to work from home, more people would use those options. One way to do this is via local zoning where projects like these can be incentivized through tax benefits or expedited permitting. California has a state law that requires dense development around public transit corridors (▫ Fig. 8.1).

8.12 Decarbonizing Aviation, Long-Range Travel, and Flying Less

Aviation is a significant part of the GHG problem, especially considering the total number of people who fly and how much they fly. It is first important to note that even in a wealthy country like the US, the media a person flies zero times per year. In other words, flying is not something most people even in the US do. It is more a mode of transportation for the top one-third population who fly every year (▫ Fig. 8.2).

The degrowth argument suggests that one solution is flying less. This means reflecting more on choices about where to go. California is leading the way in low-carbon aviation fuels. Driven by the state low-carbon fuel standard (LCFS), there are several new renewable jet fuels that are being sold in the state. One project is using waste tallow and converting it to jet fuel. Many organizations offer carbon offsets as a means to compensate for the emissions of aviation and flights. But offsets, as we learned in earlier chapters, face a number of different challenges about credibility, and in some cases, they have resulted in perverse outcomes as many critics of the Reducing Emissions from Deforestation and Forest Degradation (REDD) program have pointed out (Brown 2010) (◘ Fig. 8.3).

New aircraft are more efficient with each generation. The Airbus 350 and Boeing 787 "Dreamliner" are both the most fuel-efficient aircraft ever built. The 787 is the first aircraft made of a carbon fiber body, which reduces the weight of the aircraft (◘ Fig. 8.4).

High-speed rail is widely noted as a strategy to lessen the GHGs from long-distance transportation. There are success stories of different kinds, including in Japan, Germany, Italy, and France where there are high-speed rail systems that have been running for over 50 years now. In the US, California is building a high-speed rail system that will connect Los Angeles and San Francisco-San Jose (◘ Fig. 8.5).

Electric buses are already taking over metropolitan areas around the world. China, California, Europe, South Korea, and many other countries are adopting EV buses because they have significantly lower operating costs, including fueling and maintenance.

◘ **Fig. 8.1** The PV here is not enough to charge these EVs, but the shade is nice and every drop helps

Fig. 8.2 Micromobility solutions like e-scooters may help solve the final mile problem

Fig. 8.3 E85 fueling station in Las Vegas

Fig. 8.4 Solar power aircraft like these may find niche markets, but large passenger jets need higher-density fuels

Fig. 8.5 Taxpayer investments like Tesla have helped create jobs and clean energy devices

References

Anderson, J. M., Nidhi, K., Stanley, K. D., Sorensen, P., Samaras, C., & Oluwatola, O. A. (2014). *Autonomous vehicle technology: A guide for policymakers*. Rand Corporation, New York.

Bailis, R., & Baka, J. (2011). Constructing sustainable biofuels: Governance of the emerging biofuel economy. *Annals of the Association of American Geographers, 101*, 827–838. https://doi.org/10.1080/00045608.2011.568867.

Brown, M. L. (2010). Limiting corrupt incentives in a global REDD regime. *Ecology LQ, 37*, 237.

Energy Information Administration. (2017). Gas and diesel fuel update. https://www.eia.gov/petroleum/gasdiesel/.

Forrest, K. E., Tarroja, B., Zhang, L., Shaffer, B., & Samuelsen, S. (2016). Charging a renewable future: The impact of electric vehicle charging intelligence on energy storage requirements to meet renewable portfolio standards. *Journal of Power Sources, 336*, 63–74.

Greenblatt, J. B., & Saxena, S. (2015). Autonomous taxis could greatly reduce greenhouse-gas emissions of US light-duty vehicles. *Nature Climate Change, 5*(9), 860.

McMichael, P. (2009). The world food crisis in historical perspective. *Monthly Review, 61*(3), 32.

Mersky, A. C., & Samaras, C. (2016). Fuel economy testing of autonomous vehicles. *Transportation Research Part C: Emerging Technologies, 65*, 31–48. https://doi.org/10.1016/j.trc.2016.01.001.

National Renders Association. (2000). U.S. production of fats and oils.

Rifkin, J. (2003). *The hydrogen economy: The creation of the worldwide energy web and the redistribution of power on earth*. Penguin, New York.

Sheehan, J. J. (2009). Biofuels and the conundrum of sustainability. Current opinion in biotechnology, 20(3), 318–324.

Singer, P. W., & Friedman, A. (2014). Cybersecurity: What everyone needs to know. Oxford University Press, Oxford.

Smil, V. (2011). America's oil imports: A self-inflicted burden. *Annals of the Association of American Geographers, 101*(4), 712–716.

Sperling, D. (2018). Three Revolutions. Island Press, Washington, DC.

Tarroja, B., Zhang, L., Wifvat, V., Shaffer, B., & Samuelsen, S. (2016). Assessing the stationary energy storage equivalency of vehicle-to-grid charging battery electric vehicles. *Energy, 106*, 673–690.

Tollefson, J. (2010). Fuel and waste no bar to US nuclear growth: Report finds that plentiful fuel supplies and temporary storage will buy decades of time to develop a longer-term strategy. *Nature, 467*(7314), 376–378.

USDA. (2017). Biofuels. https://www.fas.usda.gov/commodities/biofuels.

Low Carbon Industries and the Built Environment

Contents

D. Mulvaney, *Sustainable Energy Transitions*,
https://doi.org/10.1007/978-3-030-48912-0_9

Learning Goals

By the end of this chapter, readers shall be familiar and proficient in the following:

- learn energy efficiency opportunities and other issues related to infrastructure, industries, and the built environment;
- understand opportunities for energy efficiency and trends in advanced industrial technologies; and
- explore ideas about greening industries such as cement, plastics, and chemicals that are a large portion of the overall GHG emissions problem.

Overview

Here, we investigate the underlying key concepts in energy efficiency, energy conservation, and green design that could help shape energy systems in the use of energy in the production of commodities, residential and commercial buildings, and other critical infrastructure and industries. These sectors use a lot of energy for heating and cooling, so we start with the importance of the concept of comfort, while focusing on the key concepts and technologies that will lead to these energy transitions in industries and the built environment. This includes solar hot water heaters and water-source heat pumps for hot water and radiant heat and geothermal heat pumps. Other key concepts described in this chapter include green chemistry, biomimicry, extended producer responsibility, design for the environment, and sustainable supply chains, all of which can contribute to greener built environments. The chapter reviews building and product certifications that have been used to shape green infrastructure projects. It also reviews trends in corporate social responsibility that enhances these efforts.

9.1 Energy Efficiency and Green Building

The built environment requires remarkable amounts of energy and materials. Nearly 40% of natural resources extracted from the Earth ends up in buildings. Industries use about a third of overall energy. Furthermore, many parts of the world are undergoing housing affordability crisis. Houses are already too expensive, and there are fewer affordable places to live. Sometimes greening the city results in gentrification that also impacts the communities by breaking them up (Rice et al. 2019). Buildings are one of the bigger challenges in decarbonization and sustainable energy transitions. Urban planning can be a slow-moving process, as building inventories turn over very slowly, and a major infrastructure project takes large amounts of time and resources, and the opportunities only occur at great intervals usually. There is a great new book by Hal Harvey and team (Harvey et al. 2018) that offers some really excellent insights into these areas.

Heating ventilation and air-conditioning (HVAC) systems energy use is particularly significant as it comprises over 50% of building consumption and 20% of total consumption in the US (Pérez-Lombard et al. 2008).

Nearly 40% of materials taken from the Earth end up in the built environment. Closing the loop on material flows and transitioning to a circular economy could have major implications for natural resource use.

California is recognized as one of the leading states in the US in terms of building energy efficiency. The state's first step toward increasing energy efficiency and security was the passing of the Warren-Alquist Act in 1974 and the subsequent formation of the California Energy Commission (CEC). California has made great strides toward using less energy and obtaining more of it from renewable sources. This led to a phenomenon in California known as the Rosenfeld effect, which suggests a smaller growth rate in electricity because of energy efficiency standards. The term takes its name from California Public Utilities Commission (CPUC) commissioner Art Rosenfeld who led the development of the nation's first and strongest technology-based energy efficiency standards.

Among the tools in this area of energy efficiency in buildings are standards for electronic devices and labeling schemes to promote more efficient products. The US EPA and DOE collaborate to enforce Energy Star label, which can be found on the equipment and devices—refrigerators, washing machines, and televisions—that meet the industry's best standards in terms of energy efficiency (Fig. 9.1).

Definition

A negawatt is a term coined by energy guru Amory Lovins, who used it to describe an avoided amount of power. When a 100-watt lightbulb is replaced with the same lumen LED but at 20 watts would result in 80 negawatts.

One area of innovation in recent years is improvements in heat pumps for residential heating applications. The use of energy for space heating and cooling is a major area of energy consumption in buildings and homes. Some heat pumps are air-to-air, finding heat in the outdoors on a cold day and delivering it indoors. Geothermal heat pumps are ground-to-air, meaning they find warm air in the ground and deliver it indoors. Heat pumps can also heat water. These are called air-to-water heat pumps, and they take heat from the outside and deliver it to water, so they can be used in applications where hot water is needed. An air-to-water heat pump can be used to heat a home or building too with hydronic or radiant floors. As heat pumps replace oil and gas for heat applications, that sector will incur significant GHG benefits and savings because of the improved efficiency. There are clear environmental benefits in residential homes when using solar energy for hot water heating or to help power heat pumps.

Greening buildings have clear benefits, but one major challenge is that the turnover of inventory is very slow. Retrofits or complete rebuilds only are a small

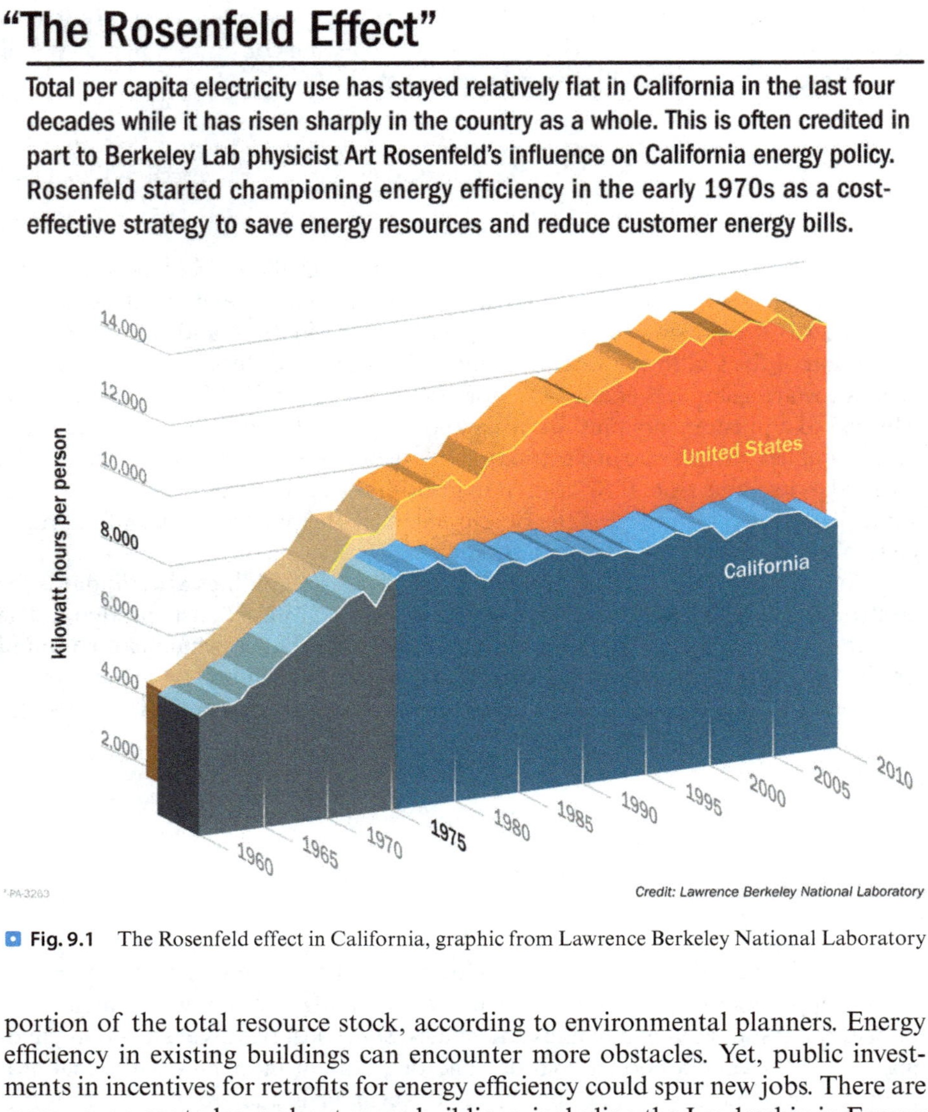

Fig. 9.1 The Rosenfeld effect in California, graphic from Lawrence Berkeley National Laboratory

portion of the total resource stock, according to environmental planners. Energy efficiency in existing buildings can encounter more obstacles. Yet, public investments in incentives for retrofits for energy efficiency could spur new jobs. There are many resources to learn about green buildings, including the Leadership in Energy and Environmental Design (LEED), Passivhaus, and the living building challenge.

9.2 Water and Wastewater Infrastructure

A major use of energy and source of GHGs in any municipality or community is related to the provision of drinking water and the disposal of wastewater. In fact, GHGs from wastewater treatment equal those from aviation and container shipping (combined all three represent about 1% of total emissions). The energy required to produce drinking water depends on where you are, and how near to the source, and the quality of the source. Some places convey water over great distances. Los Angeles gets its drinking water from hundreds of miles away in the Sierra Nevada Mountain Range, the same chain that sends water down the western slopes to the city of San Francisco. Globally, drinking water is a critical equity issue. Water in some rural households is used for their domestic tasks as well (Goff and Crow 2014).

Wastewater poses another separate set of challenges. Global sanitation is a major challenge still globally, with billions lacking access to basic sanitation and clean drinking water (See the World Health Organization for the most recent statistics). For more rural homes, there may be other means of sewage disposal and treatment, including septic tanks and leach fields. Wastewater emissions can contain the GHG methane as well. This is because anaerobic decomposition of organic matter takes place in sewers, also called as methanogenesis. Since methane is more potent of a GHG than CO_2, it is an opportunity to mitigate some GHG emissions, while generating heat or electricity on-site (Listowski et al. 2011).

Definition

Methanogenesis is a metabolic pathway where microorganisms anaerobically digest organic material and produce methane.

A small segment of drinking water pipeline in Portland, Oregon, was replaced with a small hydroelectric generator, taking advantage of the gravity in the system. These niche opportunities to use water conveyance systems could add up if effectively distributed.

9.3 Cement Production

One important area to focus on the decarbonization of industry is cement production. Cement is the second most used substance on Earth after water: it is all around us. The main input for cement is limestone, along with calcium and silicon. So the supply chains for cement begin at limestone quarries, in addition to sand, another key input. Cement is critical to construction industries for everything from residential homes, to bridges, roads and sidewalks, to commercial buildings. The technique most commonly used to produce cement in kilns not only is energy intensive but also emits CO_2 by chemical reaction. Between energy consumption and chemical reaction, energy inputs, each ton of cement results in about a ton of CO_2. In fact, estimates of global CO_2 emissions put cement at around 6% of overall GHG emissions (Rodrigues and Joekes 2011).

The following stoichiometry produces CO_2 emissions:

$$CaCO_3 + \text{Heat} \rightarrow CaO + CO_2$$

There are some more short-term changes to cement production that could reduce emissions. Fuel switching to less carbon-intensive fuels could reduce emissions. Some efforts are pursuing improved energy efficiency for this very heat-intensive energy through technological change. Wet kilns are less efficient than dry kilns, so some changes may be needed to cement-making equipment. The standard clinker-technologies for Portland cement—the most common kind—are nearing the limits of their ability to get more efficient (Klee 2010).

Material substitution can reduce some GHG emissions from cement. Calcium sulfoaluminates and calcium silicates are two critical materials to reduce the emissions-causing clinker materials. Other air pollutants from cement production include persistent organic pollutants, dioxins, heavy metals, sulfur dioxide, and particulate matter (Rodrigues and Joekes 2011).

Several cement companies across the globe are experimenting with means to capture the carbon emissions from cement production. Some companies are even attracting venture capital. Without better carbon pricing or voluntary efforts to invest in carbon capture, it is unlikely that cement industries will become low carbon on their own (Panjaitan et al. 2018). Growth of cement is still slated to continue to rise in the coming decades. We are not near peak cement, so it is critical to produce with fewer emissions in this industry as we transition forward, especially given the degree of infrastructure, homes, and buildings that are candidates for retrofits. However, cement plants are very capital intensive, so turnover can take a long time.

Other considerations with cement industries are related to occupational safety. Cement kiln dust is a widely accepted occupational hazard.

9.4 Major Alloys and Metals: Steel, Copper, Aluminum

Mining industries and their associated smelting activities are significant contributors to many environmental problems, including greenhouse gases, but also water and air quality impairing emissions, and land-use change. Iron- and steel-making are about 5% of global GHGs. Every ton of steel—because of the carbon-intensive coking process—results in 1.6–3.1 tons of CO_2e (Davis et al. 2018). Some of the fundamental processes cannot change, but increasingly, particularly in Northern Europe, there are smelters and mining operations that are mostly relying on renewable electricity purchased off-site, mostly wind-coupled with hydropower. So some of the energy impacts are a result of the energy landscapes upon which they sit.

Circular economy approaches are worth investigating on the topic of major metals and alloys. Recycled materials for metals can significantly reduce the environmental burden of metals used in products. Closing this loop results in less energy and GHGs compared to virgin metals, particularly studies of steel, copper, and aluminum. But semiconductor metal recovery also reduces environmental burdens of photovoltaics. Recycling metals is getting more complicated as different kinds of metals are brought together where historically materials were homogenous. Copper contamination from electric components in cars, for example, is complicating steel recycling. Increased plastic composites also complicate this picture. Recycled aluminum can reduce energy use by almost tenfold compared to making it from bauxite.

Some companies are experimenting with biological organisms to help collect or concentrate sought-after metals. The concept of a bio-mine is one that harvests electric e-waste of some other metal-containing waste and allows waste products to be recovered.

Yet, in many places, simply going after second-life metal will not suffice for a rapidly growing global economy that requires new metals for new energy and building technologies. Some mining activities may be required. Advances in harvesting metals from saltwater also remains a viable option, particularly if tied to desalination plants. There are even proposals to harvest key metals from geothermal or oil and gas wastewater brines (Evans 2008). Electrification or use of hydrogen in industry will be a near-term decarbonization research area.

9.5 Chemical Industries

An important area of decarbonization is the chemical industry, which is responsible for 10% of global GHG emissions, as well as numerous other materials (McKinsey and Company 2018). About two-thirds of overall energy use in industry is used by manufacturers using high temperature heat such as in making plastics and ammonia. Many chemical industries like these rely on steam and super-heated steam for heating reactor vessels and other tasks. Electrification of many of these tasks could also reduce emissions in this space. Ammonia could also be used as a fuel directly, or cracked, for its hydrogen (Davis et al. 2018) (Fig. 9.2).

Chemical industries are intimately interconnected with energy companies. Many of the feedstocks used in chemical industries are purchased from the oil and gas industries. The chemical industries are often customers of petroleum companies that sell them raw materials. One common example is ethane, which is separated by a cracker. They are also critical suppliers of materials to the photovoltaic industries, making everything from metallization pastes and solders to backsheets and encapsulants.

Principles of green chemistry are generally leading the sustainability focus in this sector. Some chemical companies are actively involved with corporate social responsibility. Much of this effort today is on ensuring the harmful materials are taken out of the plastics like halogens or brominated flame retardants. DSM, for example, is a leader in eliminating fluoropolymers and halogenated materials from plastics. The next phase will have to begin to focus on the breakdown of plastics, to ensure that we solve our plastic *and* climate pollution problems simultaneously (Fig. 9.3).

It will be important for discoveries to be made in bioplastics to replace plastics made from natural gas or ethane. There are many questions still about the quality and durability of plastics that will make it challenging to replace. But transitions away from fossil fuels will imply transitions away from plastics, and this move will help with issues related to plastics in water and oceans, the Pacific garbage patch, plastic bag pollution, and microplastics, all of which can be problematic for wildlife and natural resources like salt with increasing density of plastic pollution in the water column (Figs. 9.4, 9.5).

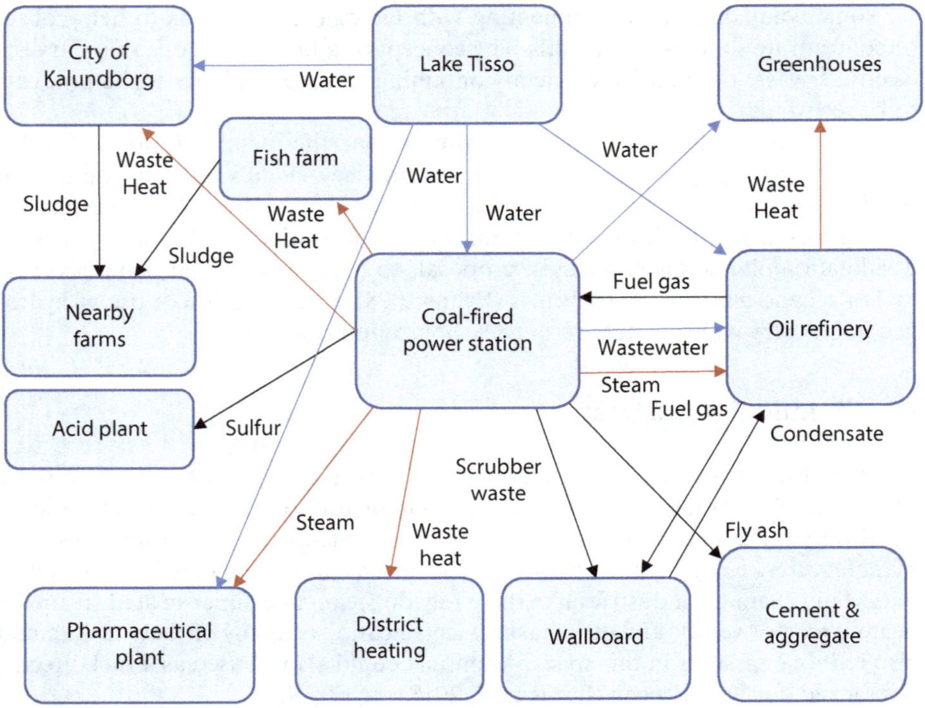

Fig. 9.2 Schematic of the industrial symbiosis at the Kalundborg complex in Denmark

Fig. 9.3 Bioplastic packaging made from corn that is a biodegradable material

Fig. 9.4 Petrochemical industries are important to renewable energy industries but also rely on extractive industries

Fig. 9.5 Pulp and paper is a promising area for circular economy

References

Davis, S. J., Lewis, N. S., Shaner, M., Aggarwal, S., Arent, D., Azevedo, I. L., & Clack, C. T. (2018). Net-zero emissions energy systems. *Science, 360*(6396), eaas9793.

Evans, R. K. (2008). *An abundance of lithium*. World Lithium: Santiago.

Goff, M., & Crow, B. (2014). What is water equity? The unfortunate consequences of a global focus on 'drinking water'. *Water International, 39*(2), 159–171.

Harvey, H., Orvis, R., & Rissman, J. (2018). *Designing climate solutions: A policy guide for low-carbon energy*. Washington, DC: Island Press.

Klee, H. (2010). The Cement Sustainability Initiative. World Business Council on Sustainable Development.

Listowski, A., Ngo, H. H., Guo, W. S., Vigneswaran, S., Shin, H. S., & Moon, H. (2011). Greenhouse gas (GHG) emissions from urban wastewater system: Future assessment framework and methodology. *Journal of Water Sustainability, 1*(1), 113–125.

McKinsey and Company. (2018). *Decarbonization of industrial sectors: the next frontier*. https://www.mckinsey.com/~/media/mckinsey/business%20functions/sustainability/our%20insights/how%20industry%20can%20move%20toward%20a%20low%20carbon%20future/decarbonization-of-industrial-sectors-the-next-frontier.ashx

Panjaitan, T. W., Dargusch, P., Aziz, A. A., & Wadley, D. (2018). Carbon management in an emissions-intensive industry in a developing economy: Cement manufacturing in Indonesia. *Case Studies in the Environment*. https://doi.org/10.1525/cse.2017.000976.

Pérez-Lombard, L., Ortiz, J., & Pout, C. (2008). A review on buildings energy consumption information. *Energy and Buildings, 40*(3), 394–398.

Rice, J. L., Cohen, D. A., Long, J., & Jurjevich, J. R. (2019). Contradictions of the climate-friendly city: New perspectives on eco-gentrification and housing justice. *International Journal of Urban and Regional Research*. https://doi.org/10.1111/1468-2427.12740.

Rodrigues, F. A., & Joekes, I. (2011). Cement industry: Sustainability, challenges and perspectives. *Environmental Chemistry Letters, 9*(2), 151–166.

Sustainable and Just Energy Strategies

Contents

D. Mulvaney, *Sustainable Energy Transitions*,
https://doi.org/10.1007/978-3-030-48912-0_10

Learning Goals

Readers of this chapter will:

- engage with dilemmas facing energy experts in the short term and long term;
- introduce integrated assessment frameworks like the food-energy-water (FEW) nexus and techno-ecological synergies to look for solutions that solve multiple issues at once;
- apply sustainability and social justice concepts to future visions of energy systems;
- utilize case studies, many of which will be firsthand accounts from the author's research; and
- think and work through the concepts in the book in a group or individual research project.

Overview

This conclusion offers an opportunity to synthesize the key features of sustainable energy transitions. Case studies of various elements of the energy transition can inform how we think about and plan for society's future energy infrastructures and resource base. Of critical importance is understanding the technical and biophysical dimensions of energy transitions, while emphasizing how problems from energy systems are ultimately social problems with a technical dimension by familiarizing students with approaches used across the social sciences.

10.1 Food-Energy-Water Nexus

Agriculture contributes about 25% of all greenhouse gas (GHG) emissions, and about 80% of that comes from products of animal agriculture such as meat, milk, and eggs. The agriculture industry has major impacts on the environment, as livestock production accounts for 70% of all agricultural land and 30% of the land surface. Water infrastructure also uses considerable amounts of energy in acquisition and conveyance and purification. Water infrastructure also works in conjunction with wastewater flows from residential and industrial production, and this consumes a lot of energy and is a major source of industrial GHG emissions.

Food, energy, and water systems are deeply interconnected. We need energy and water to grow food. Energy requires water to extract and operate. Water systems require energy to obtain, move, clean, and dispose of. These interconnections illustrate how our energy, food, and water systems are interdependent.

Definition

The food-energy-water nexus is a phrase used to describe the biophysical, natural, social, and behavioral processes that are interconnected by food, energy, and water.

How can we design interconnected systems that work to solve problems at the food-energy-water (FEW) nexus? FEW systems comprise physical processes (such as built infrastructure and new technologies for more efficient resource utilization), natural processes (such as biogeochemical and hydrological cycles), biological processes (such as agroecosystem structure and productivity), and social and behavioral processes (such as decision-making and governance). FEW systems have no particular delimitation or boundary and may appropriately be defined at a wide range of temporal and spatial scales—locally to regionally to globally.

Agriculture and forestry produce about 25% of global annual greenhouse gases. Another 25% of greenhouse gases are from electricity and heat.

Utilizing solar energy can have important synergies with several critical issues facing food production at the FEW nexus. Photovoltaic (PV) systems displace thermoelectric power generation, which consumes roughly 39% of freshwater withdrawals in the US (Finley and Seiber 2014). Freshwater use for producing food and fiber constitutes roughly 70% of the freshwater used in agriculture. In arid regions with agricultural production, the pumping and conveyance energy used to deliver water can consume large amounts of water. Watering crops in California consumes roughly 3% of the state's electricity (CEC 2005). By increasing the amount of solar energy deployment, both the GHG intensity and water intensity of food production systems can be reduced. Even modest reductions in water use can have important implications for water-intensive crop production. These problems are particularly acute where water use is high overtop critically overdrafted water basins along the California Central Coast and Central Valley (California Department of Water and Power 2015). This region is also facing issues with subsidence, and growers will soon be subject to California's 2015 Sustainable Groundwater Management Act. Roughly 50% of California's crops are exported, and with these crops go embodied water, so increased water use efficiency can lead to overall reductions in international virtual water transfers (Chen and Chen 2013).

Despite the broader climate, air, and water benefits, solar has land use impacts that can impact food and water systems in places under pressure from utility-scale project development. In California, as utility-scale solar projects shift away from controversial public lands toward agricultural land, this will result in increased pressures on prime farmland and land not suitable for agriculture, which may or may not presently provide ecosystem services. Where food production is replaced with energy production, water demand profiles will change, likely decreasing for photovoltaics, possibly increasing for biofuels.

Another important area where there may be opportunities at the FEW system nexus is with global meat production, which rose to 350 million tons in 2014 from 78 million tons in 1961. Cattle and livestock production are major sources of carbon emissions, with particularly high contributions to overall methane emissions. Methane is naturally occurring in the stomachs of cattle—referred to as enteric fermentation—and in cattle manure, where combined it emits 37% of annual methane. More than a quarter of all human-originated methane going into the atmosphere comes from raising livestock. Methane is a more potent GHG than carbon dioxide, with 23–25 times the trapping heat ability of carbon dioxide. Livestock also contribute 64% of ammonia emissions, which contribute to acid rain and acidification of land and waterways. Cattle and livestock are water-intensive also, requiring over 900 liters of water to produce a gallon of milk, and while 1.3 billion tons of grains are consumed by farm animals each year, nearly all of it is fed to cattle (FAO 2018a, b).

10.2 Sustainability and Justice Concepts for Solar Energy Futures

When controversy arose over utility-scale solar projects across the American West, policy-makers suggested finding solutions that did not require undisturbed land, leverage existing infrastructure, and protect open space. These "land-sparing" solar siting opportunities—a term introduced by a research team at the University of California, Davis, to describe brownfields, salt-contaminated lands, water bodies, particularly wastewater—are much more extensive than previously believed. The US EPA's RE-Powering America's Land initiative revealed far more lands available that could meet peak power demand many times over. Many agricultural lands and former industrial sites have already been abandoned or are unsuitable for other purposes. For example, landfills that are closed and capped have few other uses. In the US Superfund, Resource Conservation and Recovery Act (RCRA) lands and other brownfields are also usually near electricity infrastructure because the factories often took electricity at higher voltages. These brownfield-to-solar field projects also can be easy to get community buy-in as a redevelopment or as a community-led revitalization effort. But there may be issues with liability for legacy chemicals in the ground that may keep investors away from such projects, so addressing these legal ramifications may be critical to more widespread use of contaminated lands (◘ Fig. 10.1).

The impacts of solar farms on land can be significant but really depend on the existing land use and other factors such as the habitat quality. In agricultural areas, enhanced pollinator services may be something that solar farms incorporate to add additional revenue streams or to aid pollination services of adjacent lands. In agricultural areas, changes to the landscape even for solar farms can lead to loss of forage for pollinators (McCoshum and Geber 2020). Pollinators—both agricultural and native bees, for example—are undergoing population declines in some parts of the world. Adding forage for pollinators could not only keep them healthier but also increase the agricultural productivity of the landscape. There are even firms today selling honey collected at solar farms, offering land owners a secondary revenue stream from the land (◘ Fig. 10.2).

This textbook has emphasized the major land use challenges facing renewables. Finding ways to site solar but avoid land use change or offer environment benefits or ecosystem services are all options for avoiding this problem. One solution is

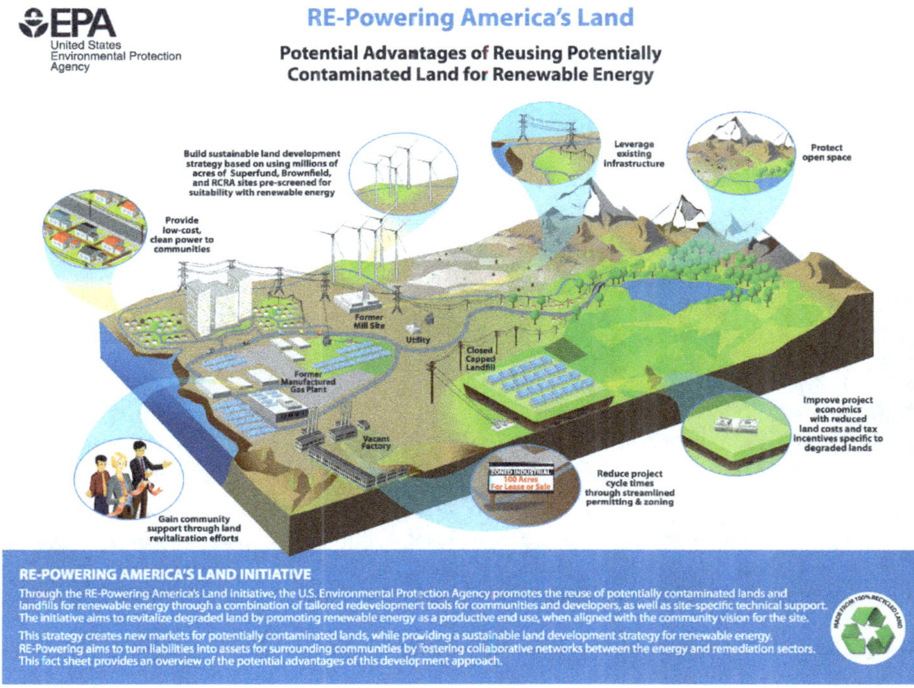

Fig. 10.1 The US EPA repowering America's land initiative offers opportunities to restore damaged land with solar farms

Fig. 10.2 Photovoltaic canopy on a parking garage at the University of Arizona

Fig. 10.3 This mesa contains cobalt, an important ingredient for batteries for energy transitions

to use photovoltaics in parking lot canopies. This keeps automobiles cooler and reduces the urban heat island effect as the photovoltaic canopies do not have the same thermal mass as pavement.

10.3 Developing Decarbonization Strategies

There are a handful of ways to approach speculation about future energy use patterns and associated environmental impact scenarios. Back-casting is an approach that selects some proposed target and constructs a scenario to transform or build an energy system to meet those requirements. Economists like to use an emissions-based impact or cost constraint and look for the most economically efficient outcome (Fig. 10.3).

Assignment 10.1 Team Project: Reduce GHGs 80% from 1990 Levels by 2050

Propose a deep decarbonization scenario where a place of your choice reduces GHGs by 80% from 1990 levels. We will collectively focus on electricity, transportation, commercial/residential, and several industry sectors.

Deliverables: (1) California Pick some country, region, or place. *WTW calculator* for the entire fleet, (2) electricity sector *fleet profile*, (3) *worksheet* detailing household emission reductions, (4) industry emission reduction *policy memo*,

(5) PowerPoint *team presentation*(s), and (6) collective *final report*.

Start a collectively shared Google Sheet in a Team Drive Find the 1990 GHG levels and final 2050 GHG targets for the place you are working on. Label the first sheet *2050 GHG targets*. Copy the sector name and category into a Google Sheet or Excel file in columns A and B. Next, put 1990 Emissions in column C and 2050 Emissions. Enter all the values for the 1990 emissions and multiply all of these values by 0.2 (20%) in column D. The title for this column should be 2050 Emissions (MMTCO2e). See this example for Agriculture. You may want to create a column between the 1990 and 2050 target, labeled 2020 forecast emissions ($MMTCO_2e$). We also need to project how much energy is needed for each of the sectors. You can collectively decide that energy use will increase or decrease by some projected amount (higher energy society, business as usual, or degrowth). Alternatively, you could evaluate three scenarios (increased, decreased, similar energy use).

This information will be helpful for preparing the introduction to the final report. Begin a shared Google doc that will have a title page, followed by a table of contents, and then an introduction.

Deep Decarbonization Final Report Outline

- Introduction
- Part 1: Transportation
- Part 2: Electricity
- Part 3: Residential and commercial
- Part 4: Industrial

Part 1: Transportation Deep Decarbonization

Complete Part 3 of the collective group WTW calculator that represents the 2050 California auto fleet.

Key questions for thinking about 2050 fleets and fuels.

- How many annual miles traveled in California?
- Annual mile reductions from efficiency, conservation, public transit?
- GHG intensity per mile traveled (WTW calculator)? Future carbon intensities?
- What proportion of engine types? How many vehicles?
- What proportion of fuel feedstocks? How many gallons/therms of fossil fuels?
- Increased electricity demand for EVs? Amount of off-grid electricity charging stations? Electrolysis?
- Should land use change be included in carbon intensity? How much land needed?

Deliverables: (1) Each person in the class is responsible for at least one row or set of rows in the group WTW calculator; add a column far to the right in which names can be entered. (2) Each group will prepare a section in Part 1 of the class final report on the considerations for biofuels, EVs, and H_2 vehicles. The outline of Part 1 should progress for each group from (a) an overview of the fuel and vehicle pathways (two pages), (b) types of technologies/resources used and extent of deployment in California in

2050 (two pages), (c) challenges and policies to drive this effort (one page), and (d) mitigations to ensure no unintended consequences (one page). Each team should have a total of six pages. We will have 24 pages overall for part of the final report.

Teams: Ethanol, biodiesel, EVs, and H_2, light-duty vehicles (LDV), heavy-duty vehicles (HDV)

Roles: Project manager, Secretary/note taker, Analyst/researcher, LDV/HDV Liaison

Part 2: Electricity

Key questions for 2050 for the future of electricity

- What is the overall electricity demand for California?
- What proportions are from wind, solar thermal, PV, geothermal, biomass, biogas, and natural gas?
- How much DG electricity? (DG PV? Fuel cells?) Off-grid? How will EVs and H_2 vehicles shape the grid?
- What proportion of electricity will come from base-load, mid-merit, peaker plants, or storage?
- How much electricity savings will come from conservation, efficiency, demand response/smart grid applications?

Deliverables: The class will produce a temporal and spatial electricity calculator by source that will estimate the space required for the future electricity sector, how much storage will be required, and so on.

Part 3: Residential/Commercial

Key Questions for 2050

- What are business-as-usual emission projections for 2050 from residential/commercial?
- How many homes can be net zero energy?
- What is the potential for solar hot water to reduce overall emissions in California?
- How can various efficiency upgrades (lighting, etc.) reduce residential and commercial energy use?

Part 4: Water, Wastewater, Agriculture, Industrial

Key Questions for 2050

- What are the GHG emissions and energy demands of the particular industry and sector?
- What are the alternative fuels, energy efficiency opportunities, and conservation impacts on these energy demands and GHG emissions?
- What kind of resources could these sectors contribute (methane from wastewater, aqueduct space for solar, Ag waste for energy)?

10.4 Critical Concepts for Sustainable Energy Strategies

Decarbonizing our economy and energy systems will require effective solutions Given the global nature of the emissions problem and the many countries involved in using GHG-emitting energy technologies, it will also require international cooperation. To move quickly, countries must be ambitious in efforts to drive deep emission cuts and stop deforestation and cap other emissions. This will involve developing better and cheaper technologies but also less-fragmented ways of implementing policies. This book aimed to synthesize the ways that many technical approaches and technological solutions require deeper investigations into the effect or impact on social systems, and planning for mitigations. Holistic approaches will be required for many of these problems, so this may require future energy system innovations and changes to designs and implementation strategies. We have seen rapid diffusions of technologies in the past. Televisions, air conditioners, cell phones, all have rapid adoption rates in very short times, so it is not out of the realm for other energy sources or technologies to be adopted as quickly.

Technology adoption is the result of complex processes. Sometimes, a technology is taken up because of some time-saving or output-enhancing feature. Turning over the global fleet of EVs, power plants, air conditioners, and water heaters might take some time. Since each of these energy-using devices has different prospects for emissions reductions, public policies would be wise to target those the most effective.

Making material systems in the industrial practice and manufacturing more circular systems will be an important element of energy transition. Green design offers opportunities to reduce material impacts on production. Extended producer responsibility (EPR) is one means to lessen the impacts of end-of-life (EOL) management problems with products after they outlive their useful life (McDonald and Pearce 2010). But there is some evidence that regulatory rules help facilitate the development of EPR in practice. Thinking through circular economy concepts can help energy planners develop systems-based connections between different processes.

Behavioral and social change is the other domain of decarbonization where there are needs to bring concepts to bear to understand where there are opportunities and obstacles to change. Here, concepts like the rebound effect, the energy efficiency gap, and energy justice could be helpful in targeting the most expeditious and equitable outcomes and strategies.

As described at the beginning, there are critical questions and debates in decarbonization that have to do with climate change itself but also other social and environmental outcomes. What might be needed are more frameworks that integrate these views.

Some people, communities, and landscapes will be more vulnerable—susceptible to harms from exposures or stressors—to energy transitions than others. Understanding how to ensure these people and places are not overly exposed to changes in human and natural systems will require careful planning and analysis.

Assignment 10.2 Key Questions for 2050 Energy Transitions

Select one of the questions below or think of your own critical question facing your community or region in its quest for more sustainable energy. Prepare some notes based on a brief research review and produce a three-page summary response to the question.

1. **How can demand response help with grid electricity management and reduce GHG emissions?**
2. **How can energy efficiency help and reduce GHG emissions?**
3. **What role can behavior play in reducing GHG emissions?**
4. **How much wind power can be responsibly developed on land?**
5. **How much wind power can be responsibly developed offshore?**
6. **How can we reduce GHG emissions from cement production?**
7. **How can we reduce GHG emissions and energy use in agriculture?**
8. **What are the obstacles to increasing distributed solar technologies?**
9. **How much natural gas can be displaced from solar hot water heating?**
10. **What is the potential for biogas at dairies?**
11. **What is the potential for biogas at wastewater treatment facilities (WWTFs)?**
12. **What is the potential for rooftop solar?**
13. **What are the opportunities and obstacles to wave power?**
14. **What are the opportunities and obstacles to geothermal power?**
15. **What should be the role of natural gas in the residential and industrial sectors?**
16. **What should be the fate of nuclear power plants that are beyond their expected life?**
17. **How will the smart grid help meet the region's climate goals?**
18. **What are the opportunities and challenges for hydropower development**
19. **How will grid operator's solve its duck curve problem?**
20. **How can electrolysis for hydrogen be incorporated into the electricity grid?**
21. **What is the potential for biomass production**
22. **What are the key policies needed to decarbonize transportation?**
23. **What are the key policies needed to decarbonize electricity?**
24. **What are the key policies needed to decarbonize the residential sector?**

There are several useful competing frameworks that deserve deeper introspection upon closer review. One design looks across the socio-energy system to evaluate recommendations to improve designs or govern the system. Hal Harvey and team have prepared an important set of policy prescriptions that could cost-effectively drive significant GHG emission reductions.

10.5 Techno-ecological Synergies

On a planet that is shared with many people and other species, it is critical to make good decisions about land energy transitions—to maximize the productivity of land and minimize the impact on land from energy infrastructure—to minimize energy sprawl. Bakshi et al. (2015) coined the idea of techno-ecological synergies to highlight the opportunities to create generative infrastructure (Fig. 10.4).

Fig. 10.4 Agrivoltaics are an example of a techno-ecological synergy. Here, rice is grown under a canopy of photovoltaics

This transition to decarbonize economies will be a global task, including efforts to establish laws to foster renewable energy development, and transition entire industries like cement and agriculture to lower carbon futures. The transition toward renewable energy, which is cost-effective today, requires a paradigm shift in thinking: the subterranean fossil fuels that power much of the world today have a small spatial yet vast environmental footprint compared to the more diffuse renewable energy resources, most of which, for now, must be collected on Earth's surface. In a world of increasing land use pressures and global assessments identifying land use change as the dominant driver (above that of climate change) of biodiversity loss, this poses a significant challenge.

Unlike other energy systems, solar energy generated from photovoltaics can be coupled with the built environment, human infrastructure, or working landscapes, including agriculture, rangelands, and aquaculture. This presents an opportunity to maximize multiple overlooked environmental benefits of solar energy deployment through techno-ecological synergies—a framework for engineering mutually beneficial relationships between technological and ecological systems.

An international group of researchers led by Dr. Rebecca R. Hernandez, a professor at the University of California, Davis, and 11 other organizations (including this book's author)—think that techno-ecological synergies represent an important framework for the sustainable production of electricity (Hernandez et al. 2019). A paper in *Nature Sustainability* provides a comprehensive examination of the opportunities to enhance the ecological and technological advantages of solar energy. Solar electricity generation coupled with sites that sequester carbon or increase ecosystem services such as improvements for pollinator habitats offered an important new framework to maximize sustainable outcomes.

Sixteen different types of solar installations with potential techno-ecological synergies were identified, including installations over previously disturbed land, over water, as distributed energy generators, in agroecological systems, and distributed throughout the electricity grid. This approach would result in land sparing of undisturbed natural areas, conserving plants and animals, including pollinator biodiversity and intact carbon sequestration cycles.

The techno-ecological synergies framework can act as a tool both for speeding the transition and for promoting smart planning of solar installations (Hernandez et al. 2019). By providing further evidence of solar energy's net benefits, the framework can be applied in regulatory and policy discussions to counter attacks by fossil fuel companies and traditional electric utilities that stand to profit from maintaining the status quo. The framework also provides an opportunity to thoughtfully maximize benefits of as many solar installations as possible. By considering the compounding climate and extinction crises, in addition to the challenges of providing electricity for a rapidly growing global population with increasing demands per capita, unintended consequences of a mass energy transition can be avoided and co-benefits embedded.

Corporations and governments are beginning to make shifts by committing to 100% clean energy and divesting from fossil fuels. These policies and practices ensure that solar energy will continue to grow rapidly in some places. But others

are dragging their feet. Implementing plans to adopt clean energy and compelling those that are slow to act to move create large questions such as how quickly the transition will happen, and where and how solar energy is developed. How an organization transitions to use of 100% clean energy—specifically including requests for additional ecosystem and agricultural benefits—is becoming increasingly important.

Across these techno-ecological synergies, the team characterized 20 specific benefits, ranging from grid resilience to land sparing to pollinator habitat, and created a matrix to communicate these synergistic outcomes. One important example the authors identified is the opportunity for solar energy on degraded lands, such as abandoned mines and brownfields. The authors found that degraded lands in the US account for almost twice the land area of California. Of this land, the most degraded sites, such as EPA Superfund sites, could produce over 1.6 million GWh per year of PV solar electricity. That's more than 35% of the total US consumption of electricity in 2015.

Center-pivot agriculture is another opportunity for techno-ecological synergy. The team estimates 21,000 km^2 or 1350 GWh could be produced in the US by utilizing these lands. This could also be directly tied to irrigation systems. Drawing on prior research from Dr. Rebecca Hernandez's lab (Hoffacker et al. 2017), the team estimates 39 TWh per year of energy potential over 104 km^2 of agricultural reservoirs in California's Central Valley. This could contribute 15% of California's annual electricity and save 0.12 km^3 of water per year.

Restoration of degraded lands can enhance ecosystem services and offers potential for carbon sequestration benefits. Research from the US Geological Survey notes that arid landscapes will be a source of emissions through 2100 because of land use change. Better land use practices could turn this source to a sink through improved conservation.

Food systems offer an incredible opportunity for solar techno-ecological synergies. "Agrivoltaic systems" or panels placed within the same land area as agricultural production have benefits including increased foraging for resources by managed and native pollinators, increased water use efficiency, and soil erosion prevention. Solar energy infrastructure can also alter microclimatic conditions that benefit overall crop production with increased water use efficiency and keep photovoltaic systems cooler and operating more efficiently. Co-siting agriculture and photovoltaic systems also result in higher yields of food and solar energy combined compared to when the food and solar energy production were separate.

Definition

Floatovoltaics are photovoltaics integrated with materials that allow them to float and be installed in lakes, reservoirs, wastewater treatment ponds, ocean bays, and other marine waterways.

On water systems, the authors examine the multiple benefits of "floatovoltaics" for panels attached to pontoons that float on water. These systems can provide 11 potential beneficial outcomes, ranging from reducing algae growth to preventing loss of water from evaporation—a benefit that is particularly important for considering solar placed over aquifers, reservoirs, and water treatment or desalination systems in water-stressed regions. The team noted prior research that found 4500 m^2 of floatovoltaics covering the entirety of the reservoir produces 425,000 kWh and saves 5,000 m^3 of water via avoided evaporation per year.

Finally, in built-up systems, we highlighted solar synergies on rooftops and other developed spaces on or near where people live and work, like parking lots. Rooftop panels can insulate buildings to improve energy savings and ultimately support human health and comfort. Panels can cool buildings and parking lots by reflecting light, potentially reducing the urban heat island effect and decreasing the need for air conditioning in hot summer months.

Emerging solar technologies, such as transparent solar panels and flexible solar for fencing, roadways, and other applications, only highlight how advances in materials science open more doors for the techno-ecological synergies. Continued innovations in solar, but also in wind, storage and energy efficiency will provide even more opportunities.

The team characterized these benefits in their paper as a promising "springboard for the integration of solar energy techno-ecological synergies into industry and society." These synergies may require their own policies, incentives, and subsidies in addition to those already in place for other clean energy technologies, including larger-scale solar.

Description	Attributes & focus	Climate solution	Authors & organizations
100% renewable energy pathways	Managerial & technical intervention, often absent social considerations (aside from broad public health benefits); nuclear cost, risks, proliferation issues weigh; some eschew biofuels for air quality or land use issues.	Non-combustion renewables or all renewables, both large scale and distributed	Jacobson et al. WWS solutions; European modeling groups–EcoFys, Neth Michael den Elzen, EU modelers—Rocky Mountain Institute, NREL, Argonne
Eco-modernists, pragmatists, and technological breakthroughs	High energy society; reliance on innovations, venture capital; energy density and intensification required for growing population; nuclear, natural gas, or CSS required; energy efficiency rebounds	Nuclear, natural gas, innovations in next generation renewables	
Small is beautiful-decentralized energy & appropriate technology	Grid defection; defense of PURPA in US	Conservation, energy efficiency, soft technologies, often anti-nuclear; micro-grids, smart grids, demand response, etc.	Lovins circa 1976
Solar transitions, solar grand plan	Using rural arid areas to site utility-scale solar power plants.	Solar and complementary technologies	Schwarzman, Zweibel, Fthenakis, and Mason 2008; Desertec
Energy access/poverty	Energy for sustainable development; pursuit of Millennium Development Goals	Appropriate technologies that help increase quality of life	Kammen
Ethanol, hydrogen, & methanol revolutions	Technocratic silver bullet solutions, entropy thinking, all three focus on harnessing constantly replenishing waste flows	H_2 from electrolysis & biofuels; EtOH from biofuels; MeOH from CH_4 & CO_2; fuel cells can be used for each	Rifkin H_2, Shah MeOH, Blume EtOH
"All of the above" and "bridge fuel" transitions discourses	Business as usual; shift away from coal; EtOH & NG bridge fuels for electricity & transport	All energy resources; need to upgrade existing fossil fuel infrastructure to be more efficient	Shell, BP, US federal lands policy
Supply-side decarbonization strategies	Technological lock-in best shaped by cutting fossil fuels cold turkey; requires stranding assets and reshaping investment patterns;	Typically aligned with 100% renewables or appropriate technology groups, although could be aligned with any non carbon group	Center for Biological Diversity; Stockholm Institute; "keep it in the ground" and fossil fuel divestment movements

Fig. 10.5 Concepts and goals of different collectives of energy solutions

Over 800,000 km^2 of degraded land are available in the US for solar development: brownfields, Superfund sites, landfills, abandoned mines, and contaminated and abandoned agricultural lands.

This framework has the hope that it will be used in cost-benefit analysis purposes of electric rate-making, resource procurement and planning, net metering, and other value-setting processes that affect distributed solar markets, and ultimately deliver both energy and environment benefits.

10.6 Moving Forward on an Energy Transition Toward Decarbonization

This book aimed to introduce readers to the critical disciplinary debates and interdisciplinary tools to draw out their own vision. Energy transitions over the rest of the next century will be preoccupied with questions of sustainability, climate change, and energy access. To best situate human civilization for energy transitions will require interdisciplinary thinking across the domains of science and technology, social sciences, and arts and humanities, for that matter. Resilience will be a critical concept to draw on for energy transitions because it represents how well a system can rebound from change (Fig. 10.5).

There will be really important questions that accompany energy transitions. As policies hasten the retirement of fossil fuel projects and infrastructure, how will these places maintain economic activity? Where will they get new jobs? How will they replace the revenues and taxes that sometimes pay for important community resources? Some have suggested that we support workers who lose employment during these changes.

Colstrip, coal, and the just transition

Colstrip, Montana, is a small town that largely depends on the revenues from a large coal-fired power plant. The town is a case study of a town dependent on a natural resource that is seeing a major decline in demand. The case study will be important to learn from as many coal mining communities face change.

How can we support communities like this?

This energy transition challenge is a deeply social one, imbued with cultural meanings and political obstacles and technical challenges that will require interdisciplinary training and thinking. Some kind of transition is already underway in some sectors. Companies, countries, and regions are committing to 100% renewable, clean, or low-carbon energy. This is a remarkable sea change from a decade, or even five years earlier. But things may also get more challenging along the way (Figs. 10.6, 10.7).

Fig. 10.6 Communities dependent on fossil fuels are routinely exposed to industrial pollution

Fig. 10.7 Communities losing big electricity generating infrastructure still have valuable transmission access

References

Bakshi, B. R., Ziv, G., & Lepech, M. D. (2015). Techno-ecological synergy: A framework for sustainable engineering. *Environmental Science & Technology, 49*(3), 1752–1760.

California Department of Water and Power. (2015).

[CEC] California Energy Commission. California's Water-Energy Relationship, CEC-700-2005-011-SF, November 2005.

Chen, Z. M., & Chen, G. Q. (2013). Virtual water accounting for the globalized world economy: National water footprint and international virtual water trade. *Ecological Indicators, 28*, 142–149.

[FAO] Food and Agriculture Organization. (2018a). *Opportunities for solar irrigation*. Rome, Italy.

[FAO] Food and Agriculture Organization. (2018b). *Water use of livestock production systems and supply chains – Guidelines for assessment*. Rome, Italy: FAO.

Finley, J. W., & Seiber, J. N. (2014). The nexus of food, energy, and water. *Journal of Agricultural and Food Chemistry, 62*(27), 6255–6262.

Hernandez, R. R., Armstrong, A., Burney, J., Ryan, G., Moore-O'Leary, K., Diédhiou, I., ManKnick, J., Mulvaney, D., & Heath, G. (2019). Techno–ecological synergies of solar energy for global sustainability. *Nature Sustainability, 2*(7), 560–568.

Hoffacker, M. K., Allen, M. F., & Hernandez, R. R. (2017). Land-sparing opportunities for solar energy development in agricultural landscapes: A case study of the Great Central Valley, CA, United States. *Environmental Science & Technology, 51*(24), 14472–14482.

McCoshum, S. M., & Geber, M. A. (2020). Land conversion for solar facilities and urban sprawl in southwest deserts causes different amounts of habitat loss for Ashmeadiella bees. *Journal of the Kansas Entomological Society, 92*(2), 468–478.

McDonald, N. C., & Pearce, J. M. (2010). Producer responsibility and recycling solar photovoltaic modules. *Energy Policy, 38*(11), 7041–7047.

Supplementary Information

D. Mulvaney, *Sustainable Energy Transitions*,
https://doi.org/10.1007/978-3-030-48912-0

Index

D

E

H

I

J

K

L

M

T

U

V

W

X

Y

CPSIA information can be obtained
at www.ICGtesting.com
Printed in the USA
LVHW080832220721
693394LV00002B/4

9 783030 489113